# Building Microservices with Micronaut®

A quick-start guide to building high-performance
reactive microservices for Java developers

**Nirmal Singh**

**Zack Dawood**

BIRMINGHAM—MUMBAI

# Building Microservices with Micronaut®

Copyright © 2021 Packt Publishing

**Group Product Manager**: Aaron Lazar

**Publishing Product Manager**: Denim Pinto

**Senior Editor**: Storm Mann

**Content Development Editor**: Tiksha Lad

**Technical Editor**: Pradeep Sahu

**Copy Editor**: Safis Editing

**Project Coordinator**: Deeksha Thakkar

**Proofreader**: Safis Editing

**Indexer**: Tejal Daruwale Soni

**Production Designer**: Vijay Kamble

First published: September 2021

Production reference: 1240821

Published by Packt Publishing Ltd.

Livery Place

35 Livery Street

Birmingham

B3 2PB, UK.

ISBN 978-1-80056-423-7

www.packt.com

*To my dearest loving wife, Rajveer, for walking the journey of writing this book with me. You are the wind beneath my wings.*

*Nirmal Singh*

*To my wife, Waheetha, and son, Aaryan, for persuading me to complete the book. Behind every author there is a supporting and loving family.*

*Zack Dawood*

# Foreword

In March of 2018, Micronaut® co-founder, Graeme Rocher, delivered a keynote address at Madrid's Greach conference in which he introduced a brand-new, open-source, full-stack, JVM software development framework. Although Graeme and the rest of the core development team at the Framework's home, Object Computing, knew they had built a groundbreaking toolset that was poised to revolutionize the way microservices and serverless applications performed, they could not have predicted the incredible amount of enthusiasm and support the Framework and the team would quickly receive following the product's launch.

By shifting framework infrastructure into a compiler feature, the Micronaut framework has significantly transformed the way developers view startup time and memory consumption when building microservices and serverless applications. It has upended the way developers approach framework design in server-side Java and rewritten the book on how frameworks should be built in the post-Java EE world.

The Framework is being used at SmartThings, StainlessAI, Minecraft, and many other organizations around the globe. It has been featured at tech conferences around the world, reached a Trial status in the 'Languages and Frameworks' section of ThoughtWorks' Technology Radar, received its own hipster in honor of JHipster's Micronaut Blueprint, and had more than 60,000 apps created on Micronaut Launch, an online tool for quickly and easily creating Micronaut applications.

It is not at all an exaggeration to say that the Micronaut framework is a hit!

In 2020, we launched the Micronaut Foundation, a not-for-profit company that oversees the software's roadmap and development, best practices and processes, repository control, documentation and support, and fundraising. Our goals? To ensure technical innovation and advancement of the Framework, to evangelize and promote the Framework as a leading technology in the JVM space, and to build and support an ecosystem of complementary documentation, functionality, and services.

The Foundation is led by a Technology Advisory Board made up of thought leaders and representatives from different sectors of the technology industry. Bringing together a group of folks with diverse experiences, priorities, and areas of expertise is a huge win for the technology and the community.

Object Computing and the Micronaut Foundation highly value the equal opportunity represented by open-source technologies; working together with members of the Micronaut community to share information, knowledge, and expertise can only further promote success for everyone. Thus, when Zack Dawood and Nirmal Singh approached us and told us they were writing a book about the Framework, we were delighted and happy to help in any way. *Building Microservices with Micronaut* is a great example of community members channeling their enthusiasm for the Micronaut framework into a resource that we know will lead to innovative and exciting solutions with the power to change the world.

Whether you've been a fan of the Micronaut framework since its inception or you're just getting started by picking up this book, you're part of a community full of people and organizations that are passionate about creating the very best technical solutions possible – developers and tech enthusiasts who embrace the idea that all of us is better than any one of us.

We are certain that you will find the Micronaut framework to be a game-changer, and with Zack and Nirmal's guidance, you will soon find yourself enjoying all the benefits the Framework offers. And where your journey takes you from there? Well, whatever you do, you can take pride in the fact that you're a pioneer in a future that runs on intelligent compilers, lightning-fast runtimes, and technical solutions limited only by your imagination.

*Foreword by The Micronaut Foundation*

# Contributors

## About the authors

**Nirmal Singh** is an entrepreneurial-spirited renaissance mind technology leader, with 10+ years of broad and unique cross-domain experience in handling complex inter-disciplinary requirements, leading high-performing software development teams, and delivering robust software solutions. He has worked in various techno-functional roles, serving health-tech, fin-tech, retail, and social commerce verticals in both product and consulting engagements. He currently leads a product engineering group at CGI to re-engineer a classic mission-critical wealth application.

*I would like to extend my gratitude to my loving wife for her patience and encouragement throughout the journey of writing this book. Without her, this would not have been possible.*

**Zack Dawood** is a Canada-based technologist, leader, author, and speaker. Zack has 16+ years of experience in IT software product development, big data, machine learning, mobile app development, blockchain, payments, Agile, enterprise architecture, and DevOps. Zack is highly certified; his key certifications include Certified Agile Leader, Certified Scrum Professional, SAFe, ITIL, Microsoft Certified Professional, and TOGAF. He has also been a **Distinguished Toastmaster** (**DTM**) since 2018. Zack also serves on the board of directors and has assumed the role of president since November 2018 for a non-profit organization. Zack writes technical blogs and contributes to Stack Overflow.

*I would like to first and foremost thank my loving and patient wife, Waheetha, and son, Aaryan, for their continued support, patience, and encouragement throughout the long process of writing this book. Thanks also to the Packt team for their support and encouragement in completing this book.*

# About the reviewer

**Patrick Pu** is a senior full-stack developer at CGI, where he focuses on building microservice architecture-based fund accounting platforms.

Patrick has worked as a senior consultant and helped many clients to build digital solutions for personal and business banking, B2B relationship management, child behavioral research, sports streaming, and digital currency exchange.

Patrick is a cat lover and blockchain enthusiast.

# Table of Contents

# Section 2: Microservices Development

## 2
## Working on Data Access

## 3
## Working on RESTful Web Services

# 4
# Securing the Microservices

# 5

## Integrating Microservices Using Event-Driven Architecture

# Section 3: Microservices Testing

# 6

## Testing Microservices

# Section 4: Microservices Deployment

## 7

## Handling Microservice Concerns

## 8

## Deploying Microservices

# Section 5:
# Microservices Maintenance

## 9
## Distributed Logging, Tracing, and Monitoring

# Section 6:
# IoT with Micronaut and Closure

## 10
## IoT with Micronaut

# 11
# Building Enterprise-Grade Microservices

## Assessment

## Other Books You May Enjoy

## Index

# Preface

Micronaut is a JVM-based framework for building lightweight, modular applications. It is a fast-growing framework designed to make creating microservices quick and easy. This book will help full stack/Java developers to build modular, high-performing, and reactive microservices-based applications using Micronaut.

## Who this book is for

This book is for developers who have been building microservices on traditional frameworks such as Spring Boot and are looking for a faster alternative. Intermediate knowledge of Java programming is required, along with a intermediate knowledge of implementing web services development in Java.

## What this book covers

*Chapter 1*, *Getting Started with Microservices Using the Micronaut Framework*, kicks off with some conceptual fundamentals on microservices and microservices design patterns. You are then introduced to the Micronaut framework and why this is the ideal framework for developing microservices. Later, you will get hands-on with the Micronaut framework by working on hello-world projects using Maven and Gradle.

*Chapter 2*, *Working on Data Access*, covers aspects of working with various kinds of database and persistence frameworks. You will begin with an object-relational mapping framework while doing a hands-on Hibernate framework and then dive into using a persistence framework (MyBatis). Finally, you will also integrate with a non-relational database (MongoDB).

*Chapter 3*, *Working on RESTful Web Services*, starts with a discussion on data transfer objects and mappers. You then dive into working with RESTful interfaces in the Micronaut framework. Later, you will learn about Micronaut's HTTP server and client APIs.

*Chapter 4, Securing Web Services*, covers various approaches in securing web endpoints in the Micronaut framework, such as session authentication, JWT, and OAuth.

*Chapter 5, Integrating Microservices Using Event-Driven Architecture*, starts with event-driven architecture and two different models for event publishing: a pull model and a push model. You then dive into event streaming and using Apache Kafka for integrating two microservices in the pet-clinic application (sample project).

*Chapter 6, Testing the Microservices*, sheds some light on various kinds of automated testing – unit testing, service testing, and integration testing, and how to employ these testing techniques in adopting a prudent automated testing policy to reduce cost and increase the robustness of microservices.

*Chapter 7, Handling the Microservices Concerns*, covers some core concerns while working on the microservices, such as distributed configuration management, documenting service APIs, service discovery, and the API gateway. Later, you also explore the mechanisms for fault tolerance in the Micronaut framework.

*Chapter 8, Deploying the Microservices*, covers the build and deployment aspects of microservices. You will kick things off by building the container artifacts using an automated tool and then leverage Docker Compose to deploy the microservices.

*Chapter 9, Distributed Logging, Tracing, and Monitoring*, throws light on implementing the observability patterns in microservices with distributed logging, distributed tracing, and distributed monitoring.

*Chapter 10, IoT with Micronaut*, jumpstarts with an introduction to IoT with Alexa, covering Alexa fundamentals and a hello-world example. Later, you will be able to integrate Micronaut with Alexa while working on the pet-clinic application.

*Chapter 11, Building Enterprise-Grade Microservices*, covers the best practices for working on the microservices and how to build and scale enterprise-grade microservices.

# To get the most out of this book

| Software/hardware covered in the book | OS requirements |
|---|---|
| H/W | CPU: 8, RAM: 16 GB, HDD: 100 GB. |
| OS | Windows, macOS X, and Linux (any). |
| Java JDK | Version 8 or above. |
| Maven | Set up Maven in your local workspace. |
| Development IDE | Any Java-based IDE can be used. |
| Git | Set up Git locally. |
| PostegreSQL | Set up PostgreSQL either as a standalone or Docker container app. |
| MongoDB | Set up MongoDB either as a standalone or Docker container app. |
| Rest client | Set up any REST client, such as Postman or Advanced REST Client. |
| Docker | Set up Docker locally. |

**If you are using the digital version of this book, we advise you to type the code yourself or access the code via the GitHub repository (link available in the next section). Doing so will help you avoid any potential errors related to the copying and pasting of code.**

# Download the example code files

You can download the example code files for this book from GitHub at `https://github.com/PacktPublishing/Building-Microservices-with-Micronaut`. In case there's an update to the code, it will be updated on the existing GitHub repository.

We also have other code bundles from our rich catalog of books and videos available at `https://github.com/PacktPublishing/`. Check them out!

# Download the color images

We also provide a PDF file that has color images of the screenshots/diagrams used in this book. You can download it here: `https://static.packt-cdn.com/downloads/9781800564237_ColorImages.pdf`

# Conventions used

There are a number of text conventions used throughout this book.

`Code in text`: Indicates code words in text, database table names, folder names, filenames, file extensions, pathnames, dummy URLs, user input, and Twitter handles. Here is an example: "By following these instructions, we added a `foo-stream` topic and added a message to this topic."

A block of code is set as follows:

```
@KafkaClient
public interface VetReviewClient {
    @Topic("vet-reviews")
    void send(@Body VetReviewDTO vetReview);
}
```

**Bold**: Indicates a new term, an important word, or words that you see on screen. For example, words in menus or dialog boxes appear in the text like this. Here is an example: "As viewed on the Kafdrop, we can verify that our event from the pet-clinic-reviews microservice is streamed out and added to the **vet-reviews** topic."

> Tips or important notes
> Appear like this.

# Get in touch

Feedback from our readers is always welcome.

**General feedback**: If you have questions about any aspect of this book, mention the book title in the subject of your message and email us at customercare@packtpub.com.

**Errata**: Although we have taken every care to ensure the accuracy of our content, mistakes do happen. If you have found a mistake in this book, we would be grateful if you would report this to us. Please visit www.packtpub.com/support/errata, selecting your book, clicking on the Errata Submission Form link, and entering the details.

**Piracy**: If you come across any illegal copies of our works in any form on the internet, we would be grateful if you would provide us with the location address or website name. Please contact us at copyright@packt.com with a link to the material.

**If you are interested in becoming an author**: If there is a topic that you have expertise in, and you are interested in either writing or contributing to a book, please visit authors.packtpub.com.

# Share Your Thoughts

Once you've read *Building Microservices with Micronaut*, we'd love to hear your thoughts! Scan the QR code below to go straight to the Amazon review page for this book and share your feedback.

https://packt.link/r/1800564236

Your review is important to us and the tech community and will help us make sure we're delivering excellent quality content.

# Section 1: Core Concepts and Basics

This section kickstarts the microservices journey in the Micronaut framework while covering some fundamentals of microservices, microservices design patterns, and why Micronaut is the ideal framework for microservices development.

This section has the following chapter:

- *Chapter 1, Getting Started with Microservices Using the Micronaut Framework*

# 1
# Getting Started with Microservices Using the Micronaut Framework

In recent times, there's been a good buzz about **microservices** and how the **microservices architecture** has been transformational in developing rapid, agile, and enterprise-grade web services to address the unique challenges and requirements of today's world. The microservices architecture has turned the page toward disciplining the standards on developing these web services. In this chapter, we will walk through the evolution of web services to microservices. We will quickly dive into some useful microservices design patterns. We will zero in on the key pitfalls in most of the traditional Java development frameworks and how their surface-level adoption to the microservices architecture has elevated performance and optimization issues. We will then explore how the Micronaut framework has addressed these performance and optimization issues in the microservices with an overhauled and ground-up approach to microservices development. Lastly, to get started with the Micronaut framework, we will set up the Micronaut CLI and work on a small hello world project.

In this chapter, we will focus on these topics in particular:

- Introducing microservices and their evolution
- Understanding microservices design patterns
- Why Micronaut is the best choice for developing microservices
- Getting started with the Micronaut framework
- Working on the hello world project in the Micronaut framework

By the end of this chapter, you will have an understanding of how web services evolved in to microservices and why traditional Java frameworks are ineffective for developing microservices as compared to the Micronaut framework. Furthermore, we will also gain the practical knowledge to start using the Micronaut framework by working on a small project in the Micronaut framework.

## Technical requirements

All the commands and technical instructions in this chapter are run on Windows 10 and mac OS X. Code examples covered in this chapter are available in the book's GitHub repository at `https://github.com/PacktPublishing/Building-Microservices-with-Micronaut/tree/master/Chapter01`.

The following tools need to be installed and set up in the development environment:

- **Java SDK**: Version 13 or above (we used Java 14).
- **Maven**: This is optional and only required if you would like to use Maven as the build system. However, we recommend having Maven set up on any development machine. Instructions to download and install Maven can be found at `https://maven.apache.org/download.cgi`.
- **Development IDE**: Based on your preferences, any Java-based IDE can be used, but for the purpose of writing this chapter, IntelliJ was used.
- **Git**: Instructions to download and install Git can be found at `https://git-scm.com/downloads`.

# Introducing microservices and their evolution

Before we thoroughly jump into introducing and defining microservices, it will be helpful to know how microservices have evolved. In the late 1960s, Alan Kay coined the term **object-oriented programming**. Though it was a definitive idea, later it birthed the four pillars for building software solutions using object-oriented programming:

- Encapsulation

- Inheritance

- Polymorphism

- Abstraction

In a short mnemonic, it's known as EIPA. Since the inception of these four pillars, the software industry has seen the rise and fall of many programming languages, frameworks, design patterns, and so on. With each such adaption and idea, thinkers and tinkerers have tried to come closer to EIPA by keeping a modular design and loosely coupled yet tightly encapsulated application components. Over the last few decades, software teams have moved away from the art of object-oriented programming toward the science of object-oriented programming by systematically adopting these key pillars. This iterative journey is the evolution of microservices.

In the late 1980s and early 1990s, almost every enterprise application was exposed as either a command line or native desktop software. Applications were tightly connected to databases and it was almost as if the end user was directly interacting with the database with the application as a thin façade in between. It was the era of monolithic applications or client/server architecture.

In the proceeding diagram, we can see how users interacted with a monolith application:

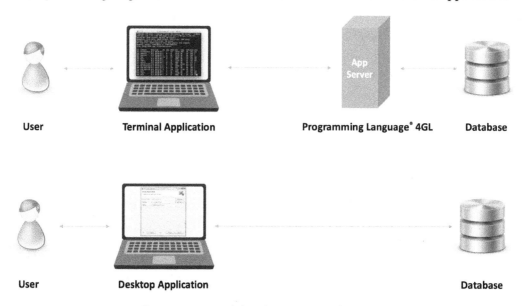

Figure 1.1 – Monolithic client/server architecture

As depicted in *Figure 1.1*, in monolithic client/server architecture, the application is tightly coupled to the database and the user interacts through a terminal façade or desktop application. In this architecture, it was painful to maintain good **service-level agreements (SLAs)**. Almost all the key non-functional factors such as scalability, high availability, fault tolerance, and flexibility underperformed or failed.

To address some of these aspects **service-oriented architecture (SOA)** came into existence. In the 2000s, SOA was formalized in the industry with the definition of some standard protocols such as **Simple Object Access Protocol (SOAP)**. **Web Services Description Language (WSDL)** was also created during this period. Web 2.0 applications were popular with **Asynchronous JavaScript And XML (AJAX)**. Enterprise service bus and messaging systems were highly used in enterprise applications. Advancements in SOA catalyzed a new paradigm of delivering software solutions to end users: **Software as a Service (SaaS)**. Instead of desktop applications and terminal clients, software solutions were delivered to end users over HTTP as hosted online services. In the proceeding diagram, we can see how users interacted with an SOA-based application:

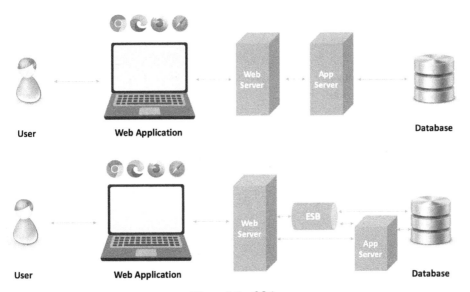

Figure 1.2 – SOA

As shown, SOA brought in some decoupling by separating the concerns between the web application, web server, and app server. App servers or **enterprise service buses (ESBs)** usually interact with the database and the user interacts with the application by accessing it on web browsers (SaaS solutions). Though SOA brought some relief, the adoption of SaaS left scalability and flexibility as key unhashed puzzles.

Post-2010, the technology world started to move much faster than it did in the previous two decades. With the introduction of containers, the cloud, big data, and machine learning, everything started moving rapidly in architecture design. It is the era of Uber, Airbnb, Netflix, and freemium/premium applications. Applications are designed for distributed computing and scalability. With the microservices architecture, the application is decomposed to loosely coupled microservices where each microservice owns its database. In the proceeding diagram, we can see how users interact with a microservices-based application:

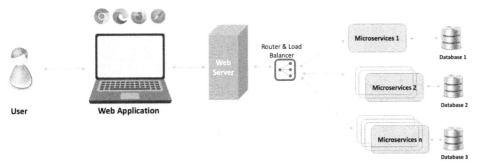

Figure 1.3 – Microservices architecture

In the previous diagram, we can see a fully realized microservices application where each microservice interacts with and owns its database. The user interacts with a single-page application through a modern browser. Any incoming requests from the web server are routed to the respective microservice. The full realization of microservices architecture is to address key factors of scalability, fault tolerance, high availability, and flexibility.

To put it simply, microservices or microservices architecture componentize an application into a collection of interacting services. Each service could be developed, tested, deployed, and maintained independently. Thus, each smaller (micro) service has its own unique life cycle. Furthermore, since each service is loosely coupled (interacting with other services using HTTP/HTTPS), we can do the following:

- Scale up or scale down (based on the service traffic).

- Address any runtime faults (boot up the service backup).

- Make new changes (change impact is limited to the service).

Therefore, through the complete realization of decoupled architecture in the microservices, we address key issues of scalability, fault tolerance, high availability, and flexibility.

So far, we have learned about microservices and their evolution and how they have been transformational in addressing the unique, rapid, and agile needs of today's world. This understanding is a good preface to realizing the potential of microservices. In the next section, we will dive into the microservices design patterns.

# Understanding microservices design patterns

To fully realize the benefits of any architecture (including the microservices architecture), an architectural approach is often backed with design patterns. Understanding these design patterns is crucial for an ideal adoption of the architecture. In the following sections, we will cover some practical and commonly used design patterns in microservices. Each pattern addresses a different aspect of the application development life cycle and our focus would be to see these design patterns from a practical usage standpoint. We will begin with decomposition design patterns.

## Decomposition design patterns

Decomposition design patterns dictate how we can componentize or decompose a big/monolithic application into smaller (micro) services. These patterns come in handy in designing a transformational architecture for any legacy monolithic application. The following are the most commonly used design patterns in decompositions.

## Decomposing by business capability

Any business capability is an instrument to make a profit. If we can enlist and categorize an application into a set of business capabilities such as inventory management, customer orders, or operations, then the application can be decomposed into microservices that are based on these business capabilities. This process to decompose is effective and recommended for small- to medium-sized applications.

## Decomposing by domains/sub-domains

If the application is an enterprise-grade and heavy application, then the previous approach may end up decomposing the application into smaller monoliths. These monoliths are smaller but monoliths nonetheless. In such cases, business modeling can help to categorize and map application functionalities into domains and sub-domains. Functionalities inside a domain/sub-domain are similar but very different from the functionalities of other domains/sub-domains. Microservices then can be designed and built around domains or sub-domains (if there are many functionalities mapped to a domain).

# Integration design patterns

Once the application is broken down into smaller (micro) services, we will need to establish cohesion among these services. Integration design patterns address such collaboration requirements. The following are the most commonly used design patterns in integrations.

## The API gateway pattern

Often upstream frontend consumers need to access microservices through a façade. This façade is called an API gateway. The API gateway design pattern serves an important purpose to keep things simple for frontend clients:

- The frontend client is not sending too many requests to microservices.
- The frontend client is not processing/aggregating too many responses (from microservices).
- At the server end, the gateway routes a request to multiple microservices, and these microservices can run in parallel.
- Before sending the final response, we can aggregate individual responses from different microservices.

## The aggregator pattern

This pattern is very similar to the aforementioned API gateway pattern. However, composite microservice is the key differential. The mandate of a composite microservice is to offload an incoming request to multiple microservices and then collaborate to create a unified response. This pattern is used when a user request is atomic from business logic standpoints, but it is processed by multiple microservices.

## The chained microservices pattern

In some scenarios, an incoming request is executed in a series of steps wherein each step could be spinning off a call to a microservice. For example, ordering an item in an online marketplace would require the following:

1. Searching for an item (inventory management service)
2. Adding an item to the cart (cart service)
3. Checking out the added item (payment service, mail service, inventory management service)

All these service calls would be synchronous. Fulfilling a user request would be an amalgamation of all these chained microservice calls.

# Data management patterns

Integrating with the persistence layer is an important aspect of any microservice-based application. Greenfield (net new) and brownfield (legacy transformation) applications may dictate their requirements in how to choose a data management pattern. The following are the most often used design patterns in data management in microservices.

## Database per service

In greenfield (net new) applications, it is ideal to have a database per service. Each service is the owner of an isolated database (relational or non-relational) and any data operation must be executed through the microservice only. Furthermore, even if any other microservice needs to perform a database operation, then it should be routed through the owner microservice.

## Shared database

In brownfield (transformational) applications, it may not be practical to decompose the database into one database per service. In such scenarios, the microservices architecture realization can be kickstarted with services sharing a common monolith database.

## Command query responsibility segregation (CQRS)

In greenfield or fully transformed applications where each microservice is an independent database owner, there might be a requirement to query data from multiple databases. The CQRS pattern stipulates to decompose an application into a command and query:

- **Command**: This part will manage any create, update, and delete requests.
- **Query**: This part will manage query requests using database views where database views can unify data from multiple schemas or data sources.

# Cross-cutting patterns

Some concerns cut across all the different aspects/layers of microservices. In the following sub-sections, we will discuss some of these concerns and patterns.

## The service discovery pattern

In a microservices-based application, each microservice may have more than one instance at runtime. Furthermore, these service instances can be added or removed at runtime based on traffic. This runtime agility can be an issue for upstream consumers in how they connect with services.

The service discovery pattern addresses this by implementing a service registry database. The service registry is a metadata store containing information such as the service name, where the service is running, and the current status of the service. Any change to the service runtime information will be updated in the service registry, for example, when a service adds a new instance or a service is down. This eases the pain for upstream consumers to connect with different microservices in the application.

## The circuit breaker pattern

In a microservices-based application, often services interact with each other by invoking endpoints. There could be a scenario where a service is calling a downstream service but the downstream service is down. Without a circuit breaker, the upstream service will keep calling the downstream service while it's down and this will keep impacting the user interaction with the application.

In the circuit breaker pattern, a downstream service call will be routed through a proxy. This proxy will timeout for a fixed interval if the downstream service is down. After the timeout expiry, the proxy will try to connect again. If successful, it will connect with the downstream service; otherwise, it will renew the timeout period. Therefore, the circuit breaker will not bombard the downstream service with unnecessary calls and it will not impact user interaction with the application.

## The log aggregation pattern

In the microservices landscape, often an incoming request will be processed by multiple services. Each service may create and log its entries. To trace any issues, it will be counter-intuitive to access these sporadic logs. By implementing a log aggregation pattern, logs could be indexed in a central place, thereby enabling easy access to all application logs. **Elasticsearch, Logstash, Kibana (ELK)** can be used to implement log aggregation.

In this section, we covered some often-used design patterns in different stages of the application life cycle. Understanding these design patterns is required to fully reap the benefits of microservices architecture. In the next section, we will dive into the Micronaut framework for developing microservices.

# Why Micronaut is the best choice for developing microservices

In the previous sections, we learned about the maturity on the architectural side of microservices. Unfortunately, on the implementation side, an overhaul shift to build/ develop microservices is not as mature as microservices architecture. To address some of these implementation challenges, many traditional Java frameworks have added small, iterative changes, but much-sought-after disruptive and overhauled changes are missing. At the core, these traditional Java frameworks have stayed almost the same since the time of monolithic services. Reflections, runtime proxies, and bulky configuration management have plagued all traditional frameworks with slower boot time and bigger memory footprints, making them unsuitable for microservices development.

Micronaut is developed from the bottom up, considering these important challenges, to organically support microservices development:

- **Dependency injection**: Micronaut uses JSR-330 `@Inject` for dependency injection. It adds the *Java inject* module to the compiler and all the annotations are processed at compile time. The compiler generates the byte code for all the classes based on the annotations that are used in their source code. This is all done at compile time. At runtime, Micronaut can instantiate the beans and read their metadata from the generated byte code and does not need to use the slow reflection API.

- **Ahead-of-time compilation**: As discussed before, one of the key contrasts is that Micronaut performs dependency injection, configuration management, and aspect-oriented programming proxying at compile time. Micronaut relies on one or more annotation processors to process the annotation metadata into **ASM**-generated (**assembly**) byte code. Furthermore, this ahead-of-time-generated byte code is further optimized by Java's **just-in-time** (**JIT**) compiler. Other frameworks use reflection and produce the annotation metadata at application boot-up. This metadata is loaded to runtime memory, therefore increasing the memory footprint. Instead of the Java Reflection API, Micronaut uses the Java annotation processor API, the Kotlin compiler plugin for annotation processors, and Groovy AST transformations for metaprogramming.

- **Faster boot-up time and lower memory consumption**: Other frameworks use Java reflections and at application boot-up, all classpaths are scanned to generate reflection metadata for each field, method, and constructor. This metadata is then used to determine and inject the required object into the application runtime. This adds significantly to boot-up time as well as runtime memory. As discussed previously, Micronaut uses ahead-of-time compilation and the Java annotation processor API to offload such work from the runtime and reduce memory requirements by not pushing unnecessary reflection metadata onto runtime memory.

- **Serverless applications support**: One of the key issues in serverless applications is boot-up time. With a huge memory footprint and slower boot-up time, traditional frameworks are not a prudent choice to develop serverless applications. Micronaut organically supports serverless application development by keeping the minimal runtime memory footprint and sub-second boot-up time. Furthermore, Micronaut natively supports commonly used cloud platforms for serverless function development.

- **Language-agnostic framework**: The Micronaut framework supports the Java, Kotlin, and Groovy programming languages. With varied support for major programming languages, developers can choose their preferred language option when considering cloud requirements. For example, for IoT requirements, Groovy could be a good option. This language-agnostic enablement makes it flexible and apt for varied requirements of mobile/web/cloud solutions.

- **Support to GraalVM**: Since Micronaut doesn't use reflections, any Micronaut-based application can be ahead-of-time compiled into a GraalVM native image. GraalVM is a universal virtual machine offered by Oracle that can run a Java application down to machine code. This increases application performance significantly. Any Micronaut application compiled to a GraalVM native image can boot up in milliseconds.

In light of the preceding key points, Micronaut stands out as a preferred framework to develop cloud-native, ultra-light, and rapid microservices. In addition, we performed a quick benchmark experiment to compare the application startup times for Micronaut versus another popular traditional framework. In the following chart, startup times are shown for both Micronaut and a traditional framework:

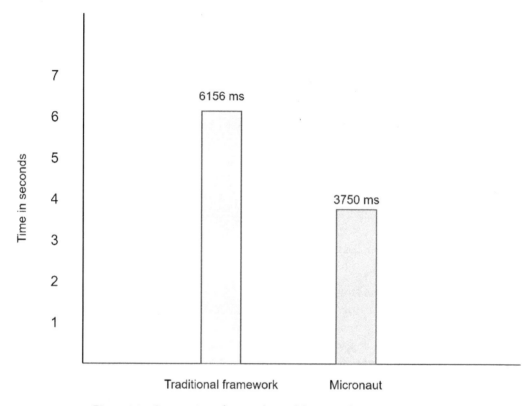

Figure 1.4 – Startup times for a traditional framework versus Micronaut

As shown in the preceding chart, the traditional framework took 6,156 milliseconds to boot up whereas Micronaut took only 3,750 milliseconds. This time difference in booting up the application is significant and sets Micronaut as a go-to framework for developing cloud-native and rapid microservices.

In the following section, we will get started with using the Micronaut framework on both Windows as well as mac OS.

# Getting started with the Micronaut framework

In order to get started with the Micronaut framework, we will begin by installing the Micronaut CLI on Mac and Windows OS.

# Installing the Micronaut CLI on mac OS

On mac OS, we can install the Micronaut CLI in a couple of ways – using SDKMAN!, Homebrew, or MacPorts. In the following sections, we will cover step-by-step instructions to install the Micronaut CLI.

## Installing Micronaut using SDKMAN!

Please follow these steps to install the Micronaut CLI using SDKMAN!:

1. Open Terminal.

2. If you don't have SDKMAN! installed, take the following steps:

   a. Type or paste the following command:

   ```
   curl -s https://get.sdkman.io | bash
   ```

   b. Next, type or paste the following command:

   ```
   source "$HOME/.sdkman/bin/sdkman-init.sh"
   ```

3. To install the Micronaut CLI, type or paste the following command:

   ```
   source sdk install micronaut
   ```

   You will observe the following interactions on Terminal while installing the Micronaut CLI:

```
                              🗋 zackdawood — -zsh — 80×24
zackdawood@zacks-mbp ~ % source "$HOME/.sdkman/bin/sdkman-init.sh"
zackdawood@zacks-mbp ~ % sdk install micronaut
==== BROADCAST ================================================================
* 2020-08-18: Jbang 0.38.0 released on SDKMAN! Checkout https://github.com/jbang
dev/jbang/releases/tag/v0.38.0. Follow @jbangdev #jbang
* 2020-08-17: Jbang 0.37.0 released on SDKMAN! Checkout https://github.com/jbang
dev/jbang/releases/tag/v0.37.0. Follow @jbangdev #jbang
* 2020-08-17: Kotlin 1.4.0 released on SDKMAN! #kotlin
===============================================================================

Downloading: micronaut 2.0.1

In progress...

################################################################### 100.0%
################################################################### 100.0%

Installing: micronaut 2.0.1
Done installing!

Setting micronaut 2.0.1 as default.
```

Figure 1.5 – Installing Micronaut CLI on mac OS using SDKMAN!

4.  If all the preceding steps execute successfully, you can verify the Micronaut CLI installation by running the following command in Terminal:

```
mn -version
```

## Installing Micronaut using Homebrew

Please follow these steps to install the Micronaut CLI using MacPorts:

1.  Open Terminal.

2.  If you don't have Homebrew installed, then take the following steps:

    a. Type or paste the following command:

    ```
    /bin/bash -c "$(curl -fsSL https://raw.githubusercontent.
    com/Homebrew/install/master/install.sh)"
    ```

    b. Next, type or paste the following command:

    ```
    brew update
    ```

3.  Type or paste the following command:

    ```
    brew install micronaut
    ```

    You will observe the following interactions on Terminal while installing the Micronaut CLI:

```
zackdawood — curl · ruby –W0 --disable=gems,did_you_mean,rubyopt /usr/local/Homebrew/Library/Homebrew/brew.rb install micron...
zackdawood@zacks-mbp ~ % brew install micronaut
==> Downloading https://homebrew.bintray.com/bottles/openjdk-14.0.1.catalina.bot
==> Downloading from https://d29vzk4ow07wi7.cloudfront.net/d44db8c5b212a36d73f11
##############                                                          21.1%
```

Figure 1.6 – Installing the Micronaut CLI on mac OS using HomeBrew

4.  If all the preceding steps execute successfully, you can verify the Micronaut CLI installation by hitting the following command in Terminal:

```
mn -version
```

## Installing Micronaut using MacPorts

Please follow these steps to install the Micronaut CLI using Homebrew:

1.  If you don't have MacPorts installed, then follow the instructions at https://www.macports.org/install.php.

2.  Open Terminal.

3.  Type or paste the following command:

```
sudo port sync
```

4.  Type or paste the following command:

```
sudo port install micronaut
```

You will observe the following interactions on Terminal while installing the Micronaut CLI:

```
                          ⚉ zackdawood — -zsh — 80×24
zackdawood@zacks-mbp ~ % sudo port install micronaut
Warning: No value for java JAVA_HOME was automatically discovered
--->   Fetching archive for micronaut
--->   Attempting to fetch micronaut-2.0.1_0.darwin_19.x86_64.tbz2 from https://p
ek.cn.packages.macports.org/macports/packages/micronaut
--->   Attempting to fetch micronaut-2.0.1_0.darwin_19.x86_64.tbz2.rmd160 from ht
tps://pek.cn.packages.macports.org/macports/packages/micronaut
--->   Installing micronaut @2.0.1_0
--->   Activating micronaut @2.0.1_0
--->   Cleaning micronaut
--->   Scanning binaries for linking errors
--->   No broken files found.
--->   No broken ports found.
zackdawood@zacks-mbp ~ %
```

Figure 1.7 – Installing the Micronaut CLI on macOS using MacPorts

5.  If all the preceding steps execute successfully, you can verify the Micronaut CLI installation by hitting the followed command in Terminal:

```
mn -version
```

# Installing the Micronaut CLI on Windows

Please follow these steps to install the Micronaut CLI on Windows:

1.  Download the Micronaut CLI binary from the Micronaut download page: https://micronaut.io/download.html.

2.  Unzip the downloaded binary file into a folder on your system. It is better to keep this in a separate folder under a root directory such as C:\Program Files\ Micronaut.

3.  Create a new system variable called MICRONAUT_HOME with the preceding directory path. Please note to add this variable under system variables (not user variables).

4. Then, update your Windows PATH environment variable. You can add a path such as %MICRONAUT_HOME%\bin.

5. Open Command Prompt or any terminal and type the following command:

```
mn
```

This will boot up the CLI for the first time by resolving any dependencies.

6. To test that the CLI is installed properly, type the following command:

```
mn - h
```

This is what the command outputs:

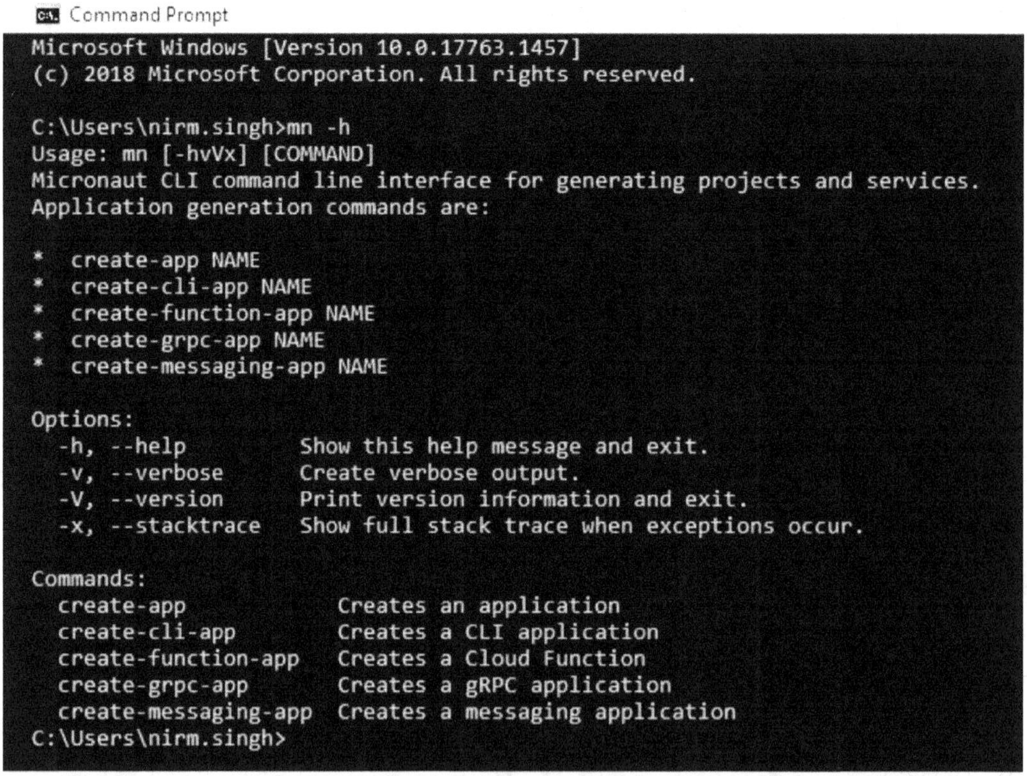

Figure 1.8 – Installing the Micronaut CLI on Windows OS

7. You should see all the **CLI** options after hitting the preceding command.

In this section, we explored different ways to install the Micronaut CLI in Windows and macOS. In order to get hands-on with the Micronaut framework, we will get started with working on a hello world project in the next section.

# Working on a hello world project in the Micronaut framework

To understand the practical aspects of using the Micronaut framework for developing a microservice, we will work with a hello world project. This will help you quickly get started with the Micronaut framework and also give you first-hand experience of how easy it is to do microservices development.

Micronaut works seamlessly with the Maven and Gradle packaging managers. We will cover one example for each using the Micronaut CLI as well as Micronaut Launch (web interface) for generating barebones projects.

## Creating a hello world project using the Micronaut CLI

Please take the following steps to create a hello world application using the Micronaut CLI:

1. Open the terminal (or Command Prompt).

2. Change the directory to your desired directory where you want to create the hello world project.

3. Type the following command:

   ```
   mn create-app hello-world-maven --build maven
   ```

4. Wait for the Micronaut CLI to finish and it will create a `hello-world-maven` project. The `create-app` command will create a boilerplate project for you with a Maven build and your system-installed Java version. It will create `Application.java` as well as a sample test class called `ApplicationTest.java`.

5. To explore your freshly created `hello-world-maven` project, open this project in your preferred IDE.

6. To run your project, run the following command in a Bash terminal:

   ```
   ./mvnw compile exec:exec
   ```

   Please note that if you are copying the project from somewhere else, then it's required to regenerate mvnw by typing the following command:

   ```
   mvn -N io.takari:maven:wrapper
   ```

7. The Maven wrapper will build and run your project on `http://localhost:8080` by default.

## Adding HelloWorldController

To create a simple endpoint, let's add a simple controller to the `hello-world-maven` project:

1.  Add a web package to our `hello-world-maven` project.

2.  Add a `HelloWorldController` Java class. It will contain a simple `hello` endpoint:

```java
@Controller("/hello")
public class HelloController {
    @Get("/")
    @Produces(MediaType.TEXT_PLAIN)
    public String helloMicronaut() {
        return "Hello, Micronaut!";
    }
}
```

    `HelloController` is accessible on the .../`hello` path. `helloMicronaut()` will generate a plain text "`Hello, Micronaut!`" message.

3.  Rerun your application and hit `http://localhost:8080/hello/` in a browser window. The server will return the following response:

Figure 1.9 – Hello, Micronaut!

By default, the application will be accessible on port `8080`, and this port can be changed in the application properties.

So far, we have worked on a hello world project using the Micronaut CLI. Next, we will explore Micronaut Launch, which is a web interface, to generate a boilerplate project.

# Creating a hello world project using Micronaut Launch

**Micronaut Launch** (`https://micronaut.io/launch/`) is an intuitive web interface that came into existence with Micronaut 2.0.1. We can use this interface to quickly generate boilerplate for different kinds of Micronaut applications (such as server applications, the CLI, serverless functions, a messaging application, and so on).
Let's quickly use this to generate a hello world application for us.

Please follow these instructions to generate the hello world project using the Micronaut Launch web interface:

1.  Open Micronaut Launch in a browser window: `https://micronaut.io/launch/`.

2.  Under **Application Type**, choose **Application**.

3.  Under **Micronaut Version**, choose **2.0.1**.

4.  For the Java version, choose **Java 14**.

5.  For **Language**, choose **Java**.

6.  Give a base package name such as `com.packtpub.micronaut`.

7.  Choose **Gradle** as the build option.

8.  Give a name to the application, such as `hello-world-gradle`.

9.  Choose **JUnit** as the testing framework

10. After you've finished choosing all the options, click on **GENERATE PROJECT**.

After choosing the preceding options and providing various inputs, the Micronaut Launch interface should look as follows:

Figure 1.10 – Using Micronaut Launch to generate a boilerplate project

Your project boilerplate source code will be generated into a zipped file. You can unarchive this zipped file into your desired directory and open it in your preferred IDE. Just like the previous example (`hello-world-maven`), we can add a basic `HelloWorldController` instance.

To run your project, run the following command in a Bash terminal:

```
gradlew.bat run
```

When the project is running, go to `http://localhost:8080/hello` and you should see the **Hello, Micronaut!** message in the browser tab.

In this section, we explored how to get started with the Micronaut framework by developing small hello world projects using the Micronaut CLI as well as the Micronaut Launch user interface. This small exercise will be a good preface for what we will cover in the next chapter.

# Summary

In this chapter, we began our journey into microservices by exploring their evolution and some useful design patterns. We covered the Micronaut framework in contrast to the traditional reflection-based Java frameworks. Essentially, Micronaut's approach to leverage ahead-of-time compilation (and not reflections) sets it apart as an ideal framework for developing microservices. To get our hands dirty, we went through setting up the Micronaut CLI on mac OS as well as Windows OS. Lastly, we worked on `hello-world-maven` and `hello-world-gradle` projects. In both projects, we added `hello` endpoints.

With the fundamentals of microservices as well as practical hello world projects covered, this chapter enhanced your knowledge of the evolution of microservices, their design patterns, and why Micronaut should be preferred for developing microservices. This foundational understanding is the bedrock for starting the adventure of microservices development in the Micronaut framework.

At the end of this chapter, we kickstarted an exciting journey into microservices development using the Micronaut CLI and Micronaut Launch. In the next chapter, we will explore how we can integrate different kinds of persistent storage and databases in the Micronaut framework.

# Questions

1.  How did web services evolve into microservices?

2.  What is a microservice?

3.  What is the microservice architecture?

4.  What are the microservices design patterns?

5.  What is Micronaut?

6.  Why should Micronaut be preferred for developing microservices?

7.  Which framework should be used to develop microservices?

8.  How do you install the Micronaut CLI on macOS?

9.  How do you install the Micronaut CLI on Windows OS?

10. How do you create a project using the Micronaut CLI?

11. How do you create a project using Micronaut Launch?

# Section 2: Microservices Development

This section will cover microservices development, including integration with the persistence layer, working on web endpoints, securing web endpoints, and integrating microservices using event-streaming. You will be doing all this while working on a microservice application in the Micronaut framework.

This section has the following chapters:

- *Chapter 2, Working on Data Access*
- *Chapter 3, Working on RESTful Web Services*
- *Chapter 4, Securing the Microservices*
- *Chapter 5, Integrating Microservices Using the Event-Driven Architecture*

# 2
# Working on Data Access

Any microservice adoption is incomplete without integrating with persistence or data storage. In this chapter, we will explore various aspects of persistence and data access in the Micronaut framework. We will begin by using an **object-relational mapping (ORM)** framework to integrate with a relational database. Then, we will dive into integrating a database using a persistence framework. Furthermore, in the end, we will see an example of integrating a NoSQL database. To cover these topics, we will work on a pet clinic application. This application will be composed of the following microservices:

- `pet-owner`: A microservice to integrate with a relational database using an ORM framework in Micronaut

- `pet-clinic`: A microservice to integrate with a relational database using a persistence framework in Micronaut

- `pet-clinic-review`: A microservice to integrate with a NoSQL database in Micronaut

By the end of this chapter, you will have good and hands-on knowledge of working with various kinds of persistence frameworks and how to integrate persistence frameworks with different kinds of databases (relational as well as NoSQL) in the Micronaut framework.

# Technical requirements

All the commands and technical instructions in this chapter are run on Windows 10 and macOS. Code examples covered in this chapter are available in the book's GitHub repository at `https://github.com/PacktPublishing/Building-Microservices-with-Micronaut/tree/master/Chapter02`.

The following tools need to be installed and set up in the development environment:

- **Java SDK**: Version 13 or above (we used Java 14).

- **Maven**: This is optional and only required if you would like to use Maven as the build system. However, we recommend having Maven set up on any development machine. Instructions to download and install Maven can be found at `https://maven.apache.org/download.cgi`.

- **Development IDE**: Based on your preferences, any Java-based IDE can be used, but for the purpose of writing this chapter, IntelliJ was used.

- **Git**: Instructions to download and install Git can be found at `https://git-scm.com/downloads`.

- **PostgreSQL**: Instructions to download and install PostgreSQL can be found at `https://www.postgresql.org/download/`.

- **MongoDB**: MongoDB Atlas provides a free online database-as-a-service with up to 512 MB storage. However, if the local database is preferred, then instructions to download and install can be found at `https://docs.mongodb.com/manual/administration/install-community/`. We used a local installation for writing this chapter.

- **Studio 3T for MongoDB**: With the MongoDB local installation, we used Studio 3T for the GUI. Instructions to download and install Studio 3T can be found at `https://studio3t.com/download/`.

# Integrating with persistence in the Micronaut framework

To exhibit integration with persistence (database) in the Micronaut framework, we will work on three different microservices within the `pet-clinic` application:

| Microservice name | Database type | Persistence framework |
|---|---|---|
| pet-owner | Relational (PostgreSQL) | Hibernate |
| pet-clinic | Relational (PostgreSQL) | MyBatis |
| pet-clinic-reviews | MongoDB | MongoDB Sync |

Figure 2.1 – Microservices in the pet-clinic application

Hibernate and MyBatis are persistence frameworks for relational databases, whereas to integrate with NoSQL (MongoDB), we will use its native synchronous driver.

In the following sections, we will cover each integration technique by doing hands-on work with the respective microservice. Each microservice (for the scope of this chapter) will be componentized into the following types of components:

- **Entity**: To encapsulate ORMs

- **Repository**: To encapsulate interaction to the underlying Hibernate framework

- **Service**: To contain any business logic as well as concierge calls to the downstream repository

- **CLI client**: To connect **create-read-update-delete** (**CRUD**) requests to the service

The following diagram depicts these components and their interaction with each other:

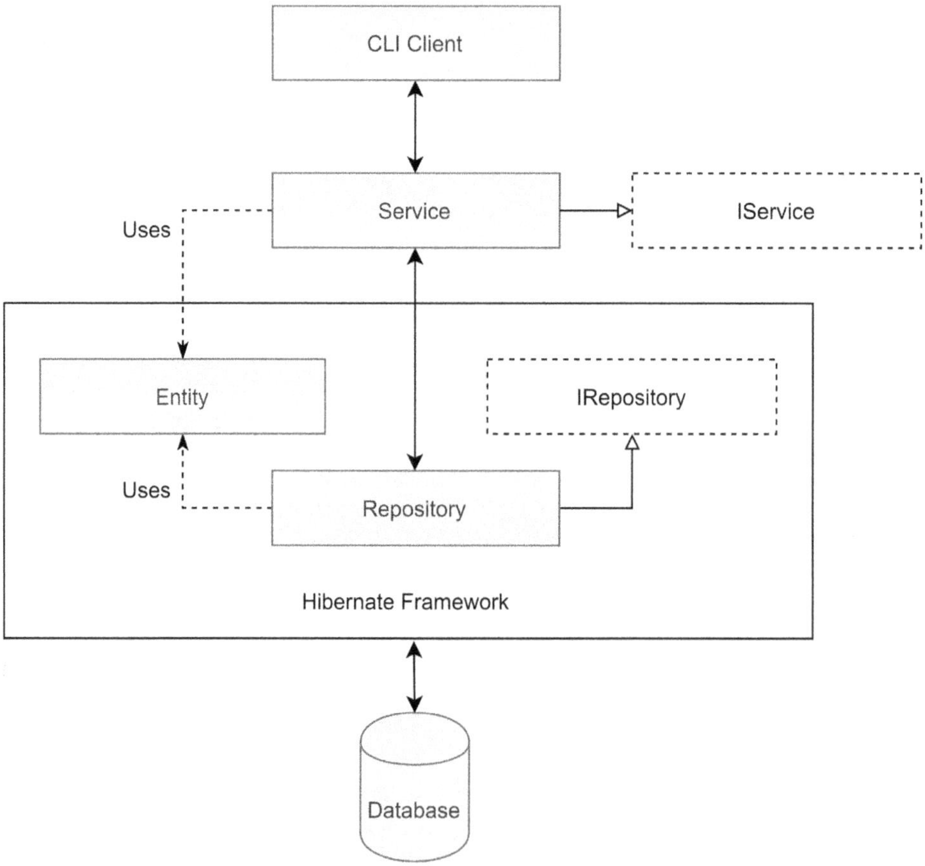

Figure 2.2 – Microservice components

We will follow the service-repository pattern to separate the concerns and decouple components within a microservice. We will cover these components in a bottom-up fashion by kickstarting with entities, then repositories, and finally services. In the next section, we will explore integrating with a relational database using an ORM framework.

# Integrating with a relational database using an ORM (Hibernate) framework

An ORM framework enables you to store, query, or manipulate data using the object-oriented paradigm. It provides an object-oriented approach to access the data from the database or, in other words, instead of using SQL, you can use Java objects to interact with the database.

In Java, as a standard specification, the **Java Persistence API (JPA)** specifies the following:

- Which Java objects ought to be persisted
- How these objects should be persisted

JPA is not a framework or tool but it dictates the standard protocol and covers the core concepts of what to persist and how to persist. Various implementing frameworks such as Hibernate and EclipseLink have adopted these JPA standards. We will be using **Hibernate** as our ORM framework.

To get hands-on with Hibernate in the Micronaut framework, we will work on the small `pet-clinic` application and, specifically for Hibernate, we will focus on the `pet-owner` microservice. The following diagram captures the schema design for the `pet-owner` microservice:

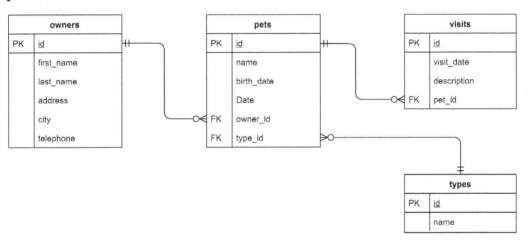

Figure 2.3 – The pet-owner schema

Essentially, in the `pet-owner` schema, one owner can have zero or more pets (of a certain type) and a pet can have zero or more vet visits. In the next section, we will get started with setting up the `pet-owner` schema.

# Generating the pet-owner schema in PostgreSQL

To generate the `pet-owner` schema, follow these instructions:

1. Download DB-SQL from `https://github.com/PacktPublishing/ Building-Microservices-with-Micronaut/blob/master/ Chapter02/micronaut-petclinic/pet-owner/src/main/ resources/db/db-sql.txt`.

2.  Open PostgreSQL's PgAdmin and open the query tool.

3.  Run the preceding SQL **section by section**. It will create a `pet-owner` user, schema, and tables.

4.  Finally, run the SQL data to ingest some dummy data into these tables.

After finishing setting up the schema, we will focus our attention on working on the Micronaut project next.

# Creating a Micronaut application for the pet-owner microservice

In order to generate boilerplate source code for the `pet-owner` microservice, we will use Micronaut Launch. Micronaut Launch is an intuitive interface to generate boilerplate and it can be accessed at `https://micronaut.io/launch/`. Once opened, this interface will look as in the following screenshot:

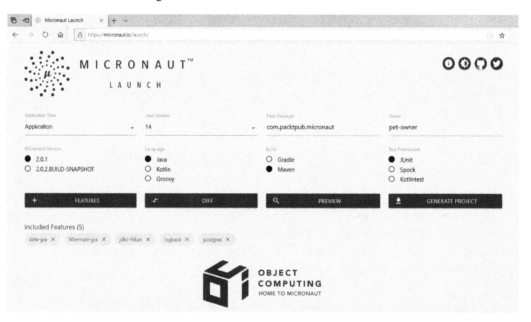

Figure 2.4 – Using Micronaut Launch to generate the pet-owner project

In Micronaut Launch, we will choose the following features (by clicking on the **FEATURES** button):

- **data-jpa**
- **hibernate-jpa**
- **jdbc-hikari**
- **logback**
- **postgres**

After specifying the aforementioned options, click on the **GENERATE PROJECT** button. A ZIP file will be downloaded onto your system. Unzip the downloaded source code into your workspace and open the project in your preferred IDE.

In the `pet-owner` microservice application, we will follow the service-repository pattern to separate the concerns and decouple components within the microservice. As discussed before, we will take a bottom-up approach in covering these components by kickstarting with entities, and then exploring repositories and finally services.

## Creating the entity classes

An entity is a **plain-old-Java-object** (**POJO**) class that is annotated with the `@Entity` annotation. An entity class definition usually contains a set of mappings for each column in the table. Therefore, an entity object instance will represent one row in the mapped table.

We will create a domain package to contain all the entity classes. We will create the `com.packtpub.micronaut.domain` package under the root package.

To map the owners table, we can define an `Owner` entity. Let's begin with mapping basic columns (skipping foreign keys or any relationships):

```
@Entity
@Table(name = "owners", schema = "petowner")
public class Owner implements Serializable {

    private static final long serialVersionUID = 1L;

    @Id
    @GeneratedValue(strategy = GenerationType.IDENTITY)
    private Long id;
```

```
    @Column(name = "first_name")
    private String firstName;

    @Column(name = "last_name")
    private String lastName;

    @Column(name = "address")
    private String address;

    @Column(name = "city")
    private String city;

    @Column(name = "telephone")
    private String telephone;
}
```

In the preceding code snippet, we are declaring an `Owner` entity class for the owners table. To map the primary key, we are using the `@Id` annotation in tandem with `@GeneratedValue`. You can generate getters and setters for the mapped columns. Similarly, we can define other entity classes: `Pet` for the pets table, `Visit` for the visits table, and `PetType` for the types table. We will take a look at defining the relationships in the next section.

# Defining relationships among entities

Using the Hibernate framework, we can define the following relationship types:

- One-to-one
- One-to-many/many-to-one
- Many-to-many

Let's take a look at each relationship type.

## Mapping a one-to-one relationship

In the `pet-owner` example, we do not have a one-to-one relationship between any entities. However, let's consider that all the address information from the owners table has been taken out into an addresses table; the resultant schema will look as in the following figure:

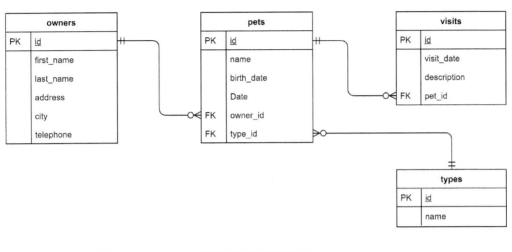

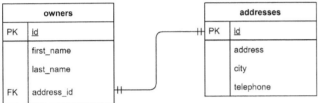

Figure 2.5 – One-to-one relationship between owners and addresses

Essentially, one owner will have one address in the preceding schema.

In the `Owner` entity, to define this relationship, we can use `@OneToOne`:

```
@Entity
@Table(name = "owners", schema = "petowner")
public class Owner implements Serializable {

    .// ...
@OneToOne(cascade = CascadeType.ALL)
    @JoinColumn(name = "address_id", referencedColumnName =
"id")
    private Address address;

// ... getters and setters
}
```

`@JoinColumn` will refer to joining the `address_id` column in the owners table.

Whereas this relationship in the `Address` entity will be defined as follows:

```java
@Entity
@Table(name = "address")
public class Address {

    @Id
    @GeneratedValue(strategy = GenerationType.IDENTITY)
    @Column(name = "id")
    private Long id;
    //...

    @OneToOne(mappedBy = "address")
    private User user;

    //... getters and setters
}
```

You may note in the `Owner` entity we have defined a one-to-one relationship using `@JoinColumn` because the owners table contains `address_id`. However, in the `Address` entity, we can simply use `mappedBy` and point to the `address` variable defined in the `Owner` entity. JPA will take care of managing this bi-directional relationship behind the scenes.

## Mapping a one-to-many/many-to-one relationship

Fortunately, in the `pet-owner` schema, we have various instances of one-to-many or many-to-one relationships (many-to-one is just a flip of a one-to-many relationship). To keep it focused, let's consider the following relationship between the owners and pets tables:

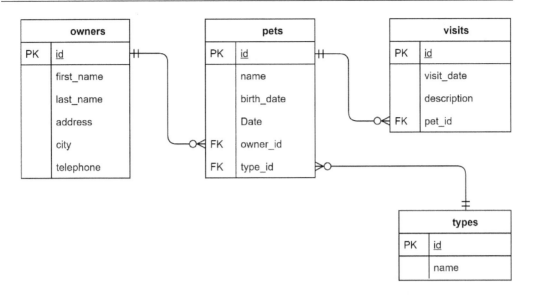

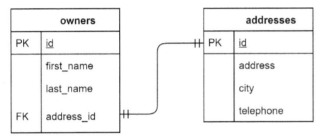

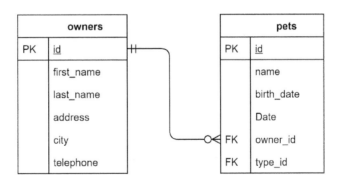

Figure 2.6 – One-to-many relationship between owners and pets

In the `Owner` entity, the preceding one-to-many relationship will be defined as follows:

```java
@Entity
@Table(name = "owners", schema = "petowner")
public class Owner implements Serializable {

    @Id
    @GeneratedValue(strategy = GenerationType.IDENTITY)
    private Long id;


    //...

    @OneToMany(mappedBy = "owner", cascade =
      CascadeType.ALL)
    private Set<Pet> pets = new HashSet<>();

    //...
```

In the `Pet` entity, this relationship will be mapped as follows:

```java
@Entity
@Table(name = "pets", schema = "petowner")
public class Pet implements Serializable {
    @Id
    @GeneratedValue(strategy = GenerationType.IDENTITY)
    private Long id;

    //...
    @ManyToOne
    @JoinColumn(name = "owner_id")
    private Owner owner;

    //...
}
```

Again, if you notice in the Pet entity we have a crisply defined relationship with Owner using @JoinColumn (because the pets table contains owner_id), whereas in the Owner entity, we simply used mappedBy = "owner". JPA will take care of defining and managing this bi-directional relationship behind the scenes.

## Mapping a many-to-many relationship

Mapping and managing a many-to-many relationship is a tad more complex. In the pet-owner schema, we don't have an instance of a many-to-many relationship, so let's assume a hypothetical relationship between two imaginary entities, Foo and Bar:

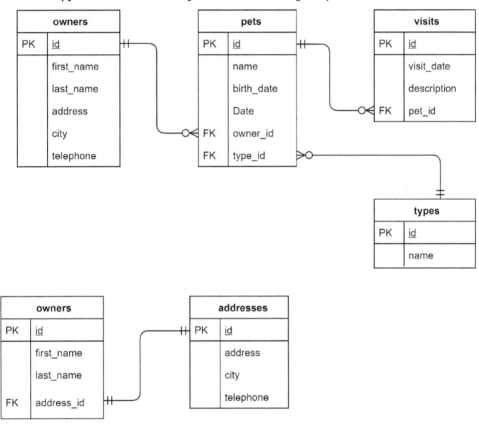

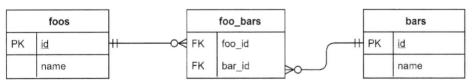

Figure 2.7 – Many-to-many relationship between foos and bars

The aforementioned many-to-many relationship will be defined as follows in the `Foo` entity:

```java
@Entity
@Table(name = "foos")
public class Foo {

    // ...

    @ManyToMany(cascade = { CascadeType.ALL })
    @JoinTable(
        name = "foo_bars",
        joinColumns = { @JoinColumn(name = "foo_id") },
        inverseJoinColumns = { @JoinColumn(name = "bar_id") }
    )
    Set<Bar> bars = new HashSet<>();

    // getters and setters
}
```

We use `@JoinTable` to map the relationship with a many-to-many table. `joinColumns` refers to the column that is owned by the entity, whereas `inverseJoinColumns` refers to the column in the co-joining entity.

The `Bar` entity will define this relationship as follows:

```java
@Entity
@Table(name = "bars")
public class Bar {

    // ...

    @ManyToMany(mappedBy = "bars")
    private Set<Foo> foos = new HashSet<>();

    // getters and setters
}
```

Just like previous examples, we have crisply defined the many-to-many relationship in the Foo entity using @JoinTable, whereas in the Bar entity we have simply used mappedBy.

Up until now, we have covered how to define different kinds of relationships in the entity classes. Next, we will divert our attention to how to create data access repositories.

## Creating data access repositories

The Hibernate framework enables us to define CRUD access to the database very intuitively. For each of the entities, we will define a repository abstract class, wherein each repository abstract class will implement JpaRepository. JpaRepository is an out-of-the-box interface defined in io.micronaut.data.jpa.repository, which further extends CrudRepository and PageableRepository to declare and define standard methods to support common CRUD operations. This reduces the syntactic sugar and frees us from defining these methods ourselves.

At the outset, we will create a new package called com.packtpub.micronaut. repository to contain all the repositories. All the repository abstract classes will look the same and here's how the OwnerRepository will look:

```
@Repository
public abstract class OwnerRepository implements
JpaRepository<Owner, Long> {

    @PersistenceContext
    private final EntityManager entityManager;

    public OwnerRepository(EntityManager entityManager) {
        this.entityManager = entityManager;
    }

    @Transactional
    public Owner mergeAndSave(Owner owner) {
        owner = entityManager.merge(owner);
        return save(owner);
    }

}
```

`OwnerRepository` is using Micronaut's standard `@Repository` annotation and it leverages `JpaRepository` to declare and define standard CRUD methods for the `Owner` entity.

Similarly, we can define these abstract classes for other entities as well – such as `Pet`, `Visit`, and `PetType` in `io.micronaut.data.jpa.repository`.

## Creating services for entities

Services will contain any business logic as well as downstream access to the repositories. We can define standard interfaces for each entity service, which will outline basic operations supported in a service.

To contain all services, we will create a package called `com.packtpub.micronaut.service`. The interface for `OwnerService` will be declared as follows:

```
public interface OwnerService {

    Owner save(Owner owner);

    Page<Owner> findAll(Pageable pageable);

    Optional<Owner> findOne(Long id);

    void delete(Long id);
}
```

The `OwnerService` interface provides an abstract declaration of all service methods. We can implement all declared methods in a concrete class:

```
@Singleton
@Transactional
public class OwnerServiceImpl implements OwnerService {

    private final Logger log =
      LoggerFactory.getLogger(OwnerServiceImpl.class);

    private final OwnerRepository ownerRepository;

    public OwnerServiceImpl(OwnerRepository
```

```java
    ownerRepository) {
        this.ownerRepository = ownerRepository;
    }


    @Override
    public Owner save(Owner owner) {
        log.debug("Request to save Owner : {}", owner);
        return ownerRepository.mergeAndSave(owner);
    }


    @Override
    @ReadOnly
    @Transactional
    public Page<Owner> findAll(Pageable pageable) {
        log.debug("Request to get all Owners");
        return ownerRepository.findAll(pageable);
    }


    @Override
    @ReadOnly
    @Transactional
    public Optional<Owner> findOne(Long id) {
        log.debug("Request to get Owner : {}", id);
        return ownerRepository.findById(id);
    }


    @Override
    public void delete(Long id) {
        log.debug("Request to delete Owner : {}", id);
        ownerRepository.deleteById(id);
    }
}
```

Service method definitions essentially delegate execution to the downstream repository. We will repeat steps to declare service interfaces and define concrete service classes for the rest of the entities – Pet, Visit, and PetType.

In the next section, we will focus our attention on creating a small command-line utility to perform common CRUD operations in the pet-owner database.

## Performing basic CRUD operations

In order to exhibit basic CRUD operations on the entities, we will create a simple utility. We can create a new package called com.packtpub.micronaut.utils to define PetOwnerCliClient:

```
@Singleton
public class PetOwnerCliClient {

    private final OwnerService ownerService;
    private final PetService petService;
    private final VisitService visitService;
    private final PetTypeService petTypeService;

    public PetOwnerCliClient(OwnerService ownerService,
                             PetService petService,
                             VisitService visitService,
                             PetTypeService petTypeService) {
        this.ownerService = ownerService;
        this.petService = petService;
        this.visitService = visitService;
        this.petTypeService = petTypeService;
    }

    // methods for performing CRUD operations...
}
```

This utility will inject all services using the constructor method. Any CRUD calls made in this utility will be executed using the injected services.

## Performing read/fetch operations

We can define a simple utility method in `PetOwnerCliClient` that can call `OwnerService` to fetch all owners. Moreover, since `Owner` has multiple pets and each pet can have multiple visits, fetching an owner will fetch pretty much everything in the `pet-owner` schema:

```
protected void performFindAll() {
    Page<Owner> pOwners =
      ownerService.findAll(Pageable.unpaged());
    … iterate through paged content
}
```

`performFindAll()` will fetch all owners and their pets (along with pet visits).

## Performing a save operation

To save an owner with a pet and a visit, we can define a method in `PetOwnerCliClient`:

```
protected Owner performSave() {
    Owner owner = initOwner();
    return ownerService.save(owner);
}


private Owner initOwner() {Owner owner = new Owner();

    owner.setFirstName("Foo");
    owner.setLastName("Bar");
    owner.setCity("Toronto");
    owner.setAddress("404 Adelaide St W");
    owner.setTelephone("647000999");

    Pet pet = new Pet();
    pet.setType(petTypeService.findAll(Pageable.unpaged()).
getContent().get(1));
    pet.setName("Baz");
    pet.setBirthDate(LocalDate.of(2010, 12, 12));
    pet.setOwner(owner);
```

```
    Visit visit = new Visit();
    visit.setVisitDate(LocalDate.now());
    visit.setDescription("Breathing issues");
    visit.setPet(pet);

    return owner;
}
```

`initOwner()` initializes an owner with a pet and pet visit, and it will be used by `performSave()` to invoke the downstream service class method to save this owner.

## Performing a delete operation

In `PetOwnerCliClient`, we will define a delete method to delete an owner (along with their pets and visits):

```
protected void performDelete(Owner owner) {
    /** delete owner pets and their visits */
    Set<Pet> pets = owner.getPets();

    if (CollectionUtils.isNotEmpty(pets)) {
        for (Pet pet : pets) {
            Set<Visit> visits = pet.getVisits();
            if (CollectionUtils.isNotEmpty(visits)) {
                for (Visit visit : visits) {
                    visitService.delete(visit.getId());
                }
            }
            petService.delete(pet.getId());
        }
    }

    ownerService.delete(owner.getId());
}
```

`performDelete()` first iterates through pets and pet visits; after deleting pet visits and pets, it will finally delegate the call to delete the owner.

# Wrapping up

To perform `PetOwnerCliClient` CRUD operations, we will add the following code logic to `Application.java`:

```java
@Singleton
public class Application {

    private final PetOwnerCliClient petOwnerCliClient;

    public Application(PetOwnerCliClient petOwnerCliClient) {
        this.petOwnerCliClient = petOwnerCliClient;
    }

    public static void main(String[] args) {
        Micronaut.run(Application.class, args);
    }

    @EventListener
    void init(StartupEvent event) {
        petOwnerCliClient.performDatabaseOperations();
    }
}
```

Then, when we run our application, the aforementioned `@EventListener` will invoke `PetOwnerCliClient` to perform database operations.

In this section, through the `pet-owner` microservice, we covered how to integrate a Micronaut-based microservice with a relational database. We also discussed how to define entities, repositories, and services, and lastly exhibited CRUD operations. In the next section, we will explore how to integrate with a relational database using another type of persistence framework (MyBatis).

# Integrating with a relational database using a persistence (MyBatis) framework

**MyBatis** is a Java persistence framework. Unlike Hibernate (an ORM framework), MyBatis does not support the direct mapping of Java objects to the database but instead maps Java methods to SQL statements.

MyBatis is commonly used in migration or transformational projects where a legacy database(s) already exists. Since a lot of tables, views, and other data objects are already defined and used in the database, it may not be an ideal scenario to refactor and normalize these table/view definitions to map them directly to Java objects (using an ORM framework). MyBatis offers an ideal way of mapping Java methods to SQL statements. These SQL statements, which manage any CRUD access thereof, are defined in an XML mapper or POJO mapper using MyBatis annotations.

Furthermore, as an ORM framework (such as Hibernate) manages child entities on its own and hides the SQL part completely, some developers prefer to have control of interacting with SQL. Therefore, MyBatis can chime in as a preferred persistence framework.

Micronaut supports integration with relational databases through MyBatis. In order to exhibit this integration, we will work on another microservice that will manage the veterinary aspect of the pet clinic application. This microservice will integrate with the following schema:

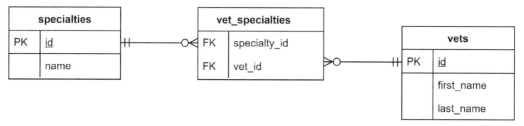

Figure 2.8 – Pet clinic schema

Essentially, one vet can have many specialties and one specialty can belong to multiple vets. In the next section, we will get started with setting up the pet-clinic schema.

## Generating the pet-clinic schema in PostgreSQL

To generate the pet-clinic schema, follow these instructions:

1.  Download DB-SQL from https://github.com/PacktPublishing/
    Building-Microservices-with-Micronaut/blob/master/
    Chapter02/micronaut-petclinic/pet-clinic/src/main/
    resources/db/db-sql.txt.

2.  Open PostgreSQL's PgAdmin and open the query tool.

3.  Run the preceding SQL **section by section**. It will create a pet-owner user, schema, and tables.

4.  Lastly, run the data SQL to ingest some dummy data into these tables.

After finishing setting up the schema, we will turn our attention to working on the Micronaut project next.

# Generating a Micronaut application for the pet-clinic microservice

In order to generate boilerplate source code for the `pet-clinic` microservice, we will use Micronaut Launch:

Figure 2.9 – Using Micronaut Launch for generating the pet-clinic project

In Micronaut Launch, we will choose the following features (by clicking on the **FEATURES** button):

- **jdbc-hikari**
- **logback**
- **postgres**

After specifying the aforementioned options, click on the **GENERATE PROJECT** button. A ZIP file will be downloaded onto your system.

Unzip the downloaded source code into your workspace and open the project in your preferred IDE. Once the project is open in the IDE, add the following dependency in pom.xml (or gradle build) for MyBatis:

```
<!-- https://mvnrepository.com/artifact/org.mybatis/mybatis -->
<dependency>
        <groupId>org.mybatis</groupId>
        <artifactId>mybatis</artifactId>
        <version>3.5.5</version>
</dependency>
```

This dependency is core to our integration with the pet-clinic schema in Micronaut.

Consistent with the service-repository pattern, we will explore MyBatis integration in a bottom-up fashion. First, we will define entities, and then repositories, and finally, we will work on services.

## Defining a MyBatis factory

To execute various SQL statements, MyBatis would need a SqlSessionFactory object at runtime. We will begin by adding a package – com.packtpub.micronaut.config. Add the following class to this newly created package:

```
@Factory
public class MybatisFactory {

    private final DataSource dataSource;

    public MybatisFactory(DataSource dataSource) {
        this.dataSource = dataSource;
    }

    @Context
    public SqlSessionFactory sqlSessionFactory() {
        TransactionFactory transactionFactory = new
        JdbcTransactionFactory();

        Environment environment = new Environment(
            "pet-clinic", transactionFactory, dataSource);
```

```
        Configuration configuration = new
          Configuration(environment);
        configuration.addMappers("com.packtpub.micronaut.
repository");

        return new
          SqlSessionFactoryBuilder().build(configuration);
    }

}
```

Using standard properties in `application.yml`, Micronaut will define a Hikari-based data source, which will be injected to define `SqlSessionFactory`. While defining the environment, you can choose any name (as we have given `pet-clinic`).

## Creating the entity classes

Similar to Hibernate entities, a MyBatis entity fundamentally defines a Java class to integrate upstream Java classes with downstream SQL interactions (defined in XML or Java mappers). However, a subtle difference is that any MyBatis entity will not contain any mapping logic.

We will add a `com.packtpub.micronaut.domain` package to contain the domain entities. Add an entity to represent `Specialty`:

```
@Introspected
public class Specialty implements Serializable {
    private static final long serialVersionUID = 1L;

    @NotNull
    private Long id;
    private String name;
    // ... getters and setters
}
```

Similarly, we can define an entity for the vets table:

```
@Introspected
public class Vet implements Serializable {
```

```
    private static final long serialVersionUID = 1L;


    @NotNull
    private Long id;


    private String firstName;


    private String lastName;


    private Set<Specialty> specialties = new HashSet<>();


    // ... getters and setters


}
```

You can note that both the entities do not have any mapping logic to map to the specialties or vets tables. In the next section, we will focus our attention on how to create data access repositories in MyBatis.

## Defining the mappers (repositories) for the entities

For the Vet and Specialty entities, we will need to define MyBatis mappers. MyBatis interacts with the downstream database using these mappers. In comparison to a typical Java application, MyBatis mappers are data access repositories. We will add a com. packtpub.micronaut.repository package to contain all these repositories.

Under this package, we will add the SpecialtyRepository interface:

```
/**
 * Mybatis mapper for {@link Specialty}.
 */
public interface SpecialtyRepository {

    @Select("SELECT * FROM petclinic.specialties")
    Collection<Specialty> findAll() throws Exception;

    @Select("SELECT * FROM petclinic.specialties WHERE id =
        #{id}")
    Specialty findById(@Param("id") Long id) throws
```

```
    Exception;

    @Select("SELECT * FROM petclinic.specialties WHERE
      UPPER(name) = #{name}")
    Specialty findByName(@Param("name") String name) throws
      Exception;

    @Select(
        {
            "INSERT INTO petclinic.specialties(id, name)\n" +
                "VALUES (COALESCE(#{id}, (select
                 nextval('petclinic.specialties_id_seq'))),
                 #{name})\n" +
                "ON CONFLICT (id)\n" +
                "DO UPDATE SET name = #{name}  \n" +
                "WHERE petclinic.specialties.id = #{id}\n" +
                "RETURNING id"
        }
    )
    @Options(flushCache = Options.FlushCachePolicy.TRUE)
    Long save(Specialty specialty) throws Exception;

    @Delete("DELETE FROM petclinic.specialties WHERE id =
     #{id}")
    void deleteById(@Param("id") Long id) throws Exception;

    @Select({
        "SELECT DISTINCT id, name FROM
          petclinic.specialties WHERE id IN(",
        "SELECT specialty_id FROM petclinic.vet_specialties
          WHERE vet_id = #{vetId}",
        ")"
    })
    Set<Specialty> findByVetId(@Param("vetId") Long vetId)
      throws Exception;
}
```

In the preceding code snippet, we can see all the actual SQL statements, which are then bound to Java methods. Therefore, whenever any upstream caller invokes any of the aforementioned methods, MyBatis will execute the corresponding mapped SQL. On the same note, we will define `VetRepository` to manage access to the vets table.

Unlike Hibernate (which provides concrete implementations for abstract repository classes), in MyBatis we will have to provide concrete implementation for the repositories. We will add the implementations to `com.packtpub.micronaut.repository. impl`.

A concrete implementation for `SpecialtyRepository` can be defined as the following:

```java
@Singleton
public class SpecialtyRepositoryImpl implements
SpecialtyRepository {


    ... SqlSessionFactory injection ...

    private SpecialtyRepository
     getSpecialtyRepository(SqlSession sqlSession) {
        return
          sqlSession.getMapper(SpecialtyRepository.class);
    }


    @Override
    public Collection<Specialty> findAll() throws Exception {
        try (SqlSession sqlSession =
          sqlSessionFactory.openSession()) {
            return
              getSpecialtyRepository(sqlSession).findAll();
        }
    }


    @Override
    public Specialty findById(Long id) throws Exception {
        try (SqlSession sqlSession =
          sqlSessionFactory.openSession()) {
            return getSpecialtyRepository
```

```java
      (sqlSession).findById(id);
    }
  }


  @Override
  public Specialty findByName(String name) throws
    Exception {
      try (SqlSession sqlSession =
       sqlSessionFactory.openSession()) {
          return getSpecialtyRepository
          (sqlSession).findByName(name);
      }
  }


  @Override
  public Long save(Specialty specialty) throws Exception {
      Long specialtyId;
      try (SqlSession sqlSession =
       sqlSessionFactory.openSession()) {
          specialtyId = getSpecialtyRepository
           (sqlSession).save(specialty);
          sqlSession.commit();
      }
      return specialtyId;
  }


  @Override
  public void deleteById(Long id) throws Exception {
      try (SqlSession sqlSession =
       sqlSessionFactory.openSession()) {
          getSpecialtyRepository(sqlSession).deleteById
           (id);
          sqlSession.commit();
      }
  }
```

```
    @Override
    public Set<Specialty> findByVetId(Long vetId) throws
     Exception {
        try (SqlSession sqlSession =
         sqlSessionFactory.openSession()) {
            return getSpecialtyRepository
             (sqlSession).findByVetId(vetId);
        }
    }
}
```

All the concrete method definitions use `SqlSessionFactory` to obtain a `SqlSession` instance. The `getSpecialtyRepository()` method will then return the MyBatis mapper using this `SqlSession` instance. Similarly, `VetRepositoryImpl` can be defined to provide concrete implementations for `VetRepository`.

In the next section, we will create upstream service classes for these repositories we just defined.

## Creating services for entities

Services will contain any business logic as well as downstream access to the preceding repositories. We can define standard interfaces for each entity service, which will outline basic operations supported in a service.

To contain all services, first, we will create a package called `com.packtpub.micronaut.service`. We can declare an interface for `SpecialtyService` to abstract our barebones structure:

```
public interface SpecialtyService {

    Specialty save(Specialty specialty) throws Exception;

    Collection<Specialty> findAll() throws Exception;

    Optional<Specialty> findOne(Long id) throws Exception;

    void delete(Long id) throws Exception;
}
```

For these barebones methods, we will need to provide concrete implementations.

We can add a package called `com.packtpub.micronaut.service.impl` under the service package. The concrete implementation for `SpecialtyService` will be defined as follows:

```
@Singleton
public class SpecialtyServiceImpl implements SpecialtyService {

    private final Logger log =
      LoggerFactory.getLogger(SpecialtyServiceImpl.class);

    private final SpecialtyRepository specialtyRepository;

    public SpecialtyServiceImpl(SpecialtyRepository
      specialtyRepository) {
        this.specialtyRepository = specialtyRepository;
    }

    @Override
    public Specialty save(Specialty specialty) throws
      Exception {
        log.debug("Request to save Specialty : {}",
          specialty);
        Long specialtyId =
          specialtyRepository.save(specialty);
        return specialtyRepository.findById(specialtyId);
    }

    @Override
    public Collection<Specialty> findAll() throws Exception {
        log.debug("Request to get all Specialties");
        return specialtyRepository.findAll();
    }

    @Override
    public Optional<Specialty> findOne(Long id) throws
      Exception {
```

```
        log.debug("Request to get Specialty : {}", id);
        return Optional.ofNullable
          (specialtyRepository.findById(id));
    }


    @Override
    public void delete(Long id) throws Exception {
        log.debug("Request to delete Specialty : {}", id);
        specialtyRepository.deleteById(id);
    }
}
```

Concrete service methods are very simple in nature and concierge any calls to downstream repository methods. We will add a similar service interface and concrete implementation for Vet as VetService and VetServiceImpl.

In the next section, we will focus our attention on creating a small command-line utility to perform common CRUD operations in the pet-clinic database.

## Performing basic CRUD operations

To use services and repositories defined in previous sections and to perform basic CRUD operations on the entities, we can create a simple utility. We can create a new package called com.packtpub.micronaut.utils to define PetClinicCliClient:

```
@Singleton
public class PetClinicCliClient {

    private final VetService vetService;
    private final SpecialtyService specialtyService;

    public PetClinicCliClient(VetService vetService,
                              SpecialtyService
                              specialtyService) {
        this.vetService = vetService;
        this.specialtyService = specialtyService;
    }

    // ...
}
```

`PetClinicCliClient` injects `VetService` and `SpecialtyService` using the constructor method. These services will then be used to perform various database operations on the `vets`, `specialties`, and `vet_specialties` tables.

## Performing read/fetch operations

We can define a simple utility method in `PetClinicCliClient` that can call `VetService` to fetch all vets. Moreover, since a vet can have multiple specialties, fetching a vet will fetch from the specialties table as well:

```
protected void performFindAll() {
    List<Vet> vets;
    try {
        vets = (List<Vet>) vetService.findAll();
        … iterate on vets
    } catch (Exception e) {
        log.error("Exception: {}", e.toString());
    }
}
```

The `performFindAll()` method fetches all vets along with their specialties from the database.

## Performing a save operation

We will save a vet with a specialty in `PetClinicCliClient`:

```
protected Vet performSave() {
    Vet vet = initVet();
    Vet savedVet = null;
    try {
        savedVet = vetService.save(vet);
    } catch (Exception e) {
        log.error("Exception: {}", e.toString());
    }
    return savedVet;
}

private Vet initVet() {
    Vet vet = new Vet();
```

```
        vet.setFirstName("Foo");
        vet.setLastName("Bar");


        Specialty specialty = new Specialty();
        specialty.setName("Baz");


        vet.getSpecialties().add(specialty);


        return vet;
    }
```

The initVet() method initializes a new vet with a specialty, which is then used by the performSave() method to persist this object to database tables.

### Performing a delete operation

In PetClinicCliClient, we will define a delete method to delete a vet (along with their specialties):

```
protected void performDelete(Vet vet) {
    try {
        vetService.delete(vet.getId());
    } catch (Exception e) {
        log.error("Exception: {}", e.toString());
    }
}
```

The performDelete() method delegates a call to VetService, which then calls the repository to finally delete the vet from the database.

## Wrapping up

To perform PetClinicCliClient CRUD operations, we will add the following code logic to Application.java:

```
@Singleton
public class Application {


    private final PetClinicCliClient petClinicCliClient;
```

```
public Application(PetClinicCliClient
  petClinicCliClient) {
    this.petClinicCliClient = petClinicCliClient;
}

public static void main(String[] args) {
    Micronaut.run(Application.class, args);
}

@EventListener
void init(StartupEvent event) {
    petClinicCliClient.performDatabaseOperations();
}
}
```

When we run our application, the aforementioned @EventListener will invoke PetClinicCliClient to perform various database operations.

In this section, we integrated with a relational database using MyBatis. We covered how to define entities, repositories, and services, and lastly, we exhibited basic CRUD operations through a utility. In the next section, we will explore integration with a NoSQL database in the Micronaut framework.

# Integrating with a NoSQL database (MongoDB)

**MongoDB** is a document-based database and it stores the data in JSON or BSON format. Data is stored in key-value pairs, similar to a JSON object. MongoDB is engineered in a scale-out fashion and it is recommended to use it when the volume and structure of data are agile and growing very rapidly. There are a few key terms in contrast to a relational database:

- **Database**: A database in MongoDB is much similar to a database in a relational database.

- **Table**: A collection (of documents) is much similar to a table in a relational database.

- **Row**: A BSON or JSON document will be a close analogy to a row in a relational database.

In order to do hands-on work, we will continue with the `pet-clinic` application and add a new microservice, that is, `pet-clinic-reviews`. This microservice will be responsible for managing vet reviews. As reviews could grow rapidly and a schema to store a review could change, we will prefer to store this data in MongoDB:

```
{
    "_id" : ObjectId("5f485523d0cfa84e00963fe4"),
    "reviewId" : "0ee19b1c-ec8e-11ea-adc1-0242ac120002",
    "rating" : 4.0,
    "comment" : "Good vet in the area",
    "vetId" : 1.0,
    "dateAdded" : ISODate("2020-08-28T00:51:47.922+0000")
}
```

The preceding JSON document depicts a review stored in MongoDB.

In the next section, we will get started with setting up the `pet-clinic-reviews` schema in MongoDB.

## Creating a vet-reviews collection in MongoDB

For the `pet-clinic-reviews` microservice, we will create a database and collection in the MongoDB instance:

1.  Open Studio 3T or any other preferred GUI for MongoDB.

2.  Connect to your localhost instance or any other preferred instance (such as MongoDB Atlas).

3.  Under **Databases**, add a new database: `pet-clinic-reviews`.

4.  In the aforementioned database, we will create a new collection: `vet-reviews`.

5.  To ingest some dummy data into this collection, use the import option to import data (CSV or JSON) from `https://github.com/PacktPublishing/Building-Microservices-with-Micronaut/tree/master/Chapter02/micronaut-petclinic/pet-clinic-reviews/src/main/resources/db`.

After finishing setting up the schema, we will turn our attention to working on the Micronaut project next.

# Generating a Micronaut application for the pet-clinic-reviews microservice

In order to generate boilerplate source code for the `pet-clinic-reviews` microservice, we will use Micronaut Launch. It is an intuitive interface to generate boilerplate:

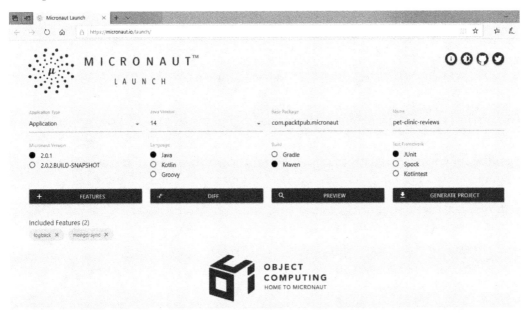

Figure 2.10 – Using Micronaut Launch to generate the pet-clinic-reviews project

In Micronaut Launch, we will choose the following features (by clicking on the **FEATURES** button):

- **mongodb-sync** (synchronous access to MongoDB)
- **logback**

After specifying the aforementioned options, click on the **GENERATE PROJECT** button. A ZIP file will be downloaded onto your system. Unzip the downloaded source code into your workspace and open the project in your preferred IDE.

In the following sections, we will dive deeper into how to integrate the `pet-clinic-reviews` microservice with MongoDB.

## Configuring MongoDB in Micronaut

In the freshly created `pet-clinic-reviews` project, we would need to update the URI in `application.yml` to point to the correct instance of MongoDB. Furthermore, we will define two new custom properties – `databaseName` and `collectionName`:

```yaml
micronaut:
  application:
    name: Pet Clinic Reviews
mongodb:
  uri: mongodb://mongodb:mongodb@localhost:27017/pet-clinic-
reviews
  databaseName: pet-clinic-reviews
  collectionName: vet-reviews
```

To use the aforementioned database and collection across the project, we can define a configuration class in a new package – `com.packtpub.micronaut.config`:

```java
public class MongoConfiguration {

    @Value("${mongodb.databaseName}")
    private String databaseName;

    @Value("${mongodb.collectionName}")
    private String collectionName;

    public String getDatabaseName() {
        return databaseName;
    }

    public String getCollectionName() {
        return collectionName;
    }
}
```

This will enable easy access to both the `databaseName` and `collectionName` properties across the project and we will not need to use `@Value`. Besides, since these properties are read-only, we will only define getters for these properties.

Next, we will create an entity class to define the vet review object.

# Creating the entity class

At the root level, add a new package called `com.packtpub.micronaut.domain`. This package will contain the domain/entity for the vet review. We can define the following POJO to represent a vet review:

```java
public class VetReview {

    private String reviewId;
    private Long vetId;
    private Double rating;
    private String comment;
    private LocalDate dateAdded;


    @BsonCreator
    @JsonCreator
    public VetReview(
            @JsonProperty("reviewId")
            @BsonProperty("reviewId") String reviewId,
            @JsonProperty("vetId")
            @BsonProperty("vetId") Long vetId,
            @JsonProperty("rating")
            @BsonProperty("rating") Double rating,
            @JsonProperty("comment")
            @BsonProperty("comment") String comment,
            @JsonProperty("dateAdded")
            @BsonProperty("dateAdded") LocalDate dateAdded) {
        this.reviewId = reviewId;
        this.vetId = vetId;
        this.rating = rating;
        this.comment = comment;
        this.dateAdded = dateAdded;
    }
    // ... getters and setters
}
```

@BsonCreator and @JsonCreator piggyback on the class constructor to map an object of VetReview to the corresponding BSON or JSON document. This will come in handy while fetching or storing a document from MongoDB and therefore mapping it to a Java object.

Next, we will focus our attention on how to create the data access repository for VetReview.

## Creating a data access repository

To manage access to the vet-reviews collection, we will create a repository. We can add com.packtpub.micronaut.repository to contain this repository class.

To abstract out all the methods that the repository will expose, we can declare an interface:

```java
public interface VetReviewRepository {

    List<VetReview> findAll();

    VetReview findByReviewId(String reviewId);

    VetReview save(VetReview vetReview);

    void delete(String reviewId);
}
```

VetReviewRepository outlines essential operations supported in this repository.

We will need to provide a concrete implementation for the VetReviewRepository interface in the com.packtpub.micronaut.repository.impl package:

```java
@Singleton
public class VetReviewRepositoryImpl implements
VetReviewRepository {

    private final MongoClient mongoClient;

    private final MongoConfiguration mongoConfiguration;

    public VetReviewRepositoryImpl(MongoClient mongoClient,
      MongoConfiguration mongoConfiguration) {
```

```java
        this.mongoClient = mongoClient;
        this.mongoConfiguration = mongoConfiguration;
    }

    @Override
    public List<VetReview> findAll() {
        List<VetReview> vetReviews = new ArrayList<>();
        getCollection().find().forEach(vetReview -> {
            vetReviews.add(vetReview);
        });
        return vetReviews;
    }

    @Override
    public VetReview findByReviewId(String reviewId) {
        return getCollection().find(eq("reviewId",
         reviewId)).first();
    }

    @Override
    public VetReview save(VetReview vetReview) {
        getCollection().insertOne(vetReview).
         getInsertedId();
        return findByReviewId(vetReview.getReviewId());
    }

    @Override
    public void delete(String reviewId) {
        getCollection().deleteOne(eq("reviewId",
         reviewId));
    }

    private MongoCollection<VetReview> getCollection() {
        return mongoClient
                .getDatabase(mongoConfiguration.
getDatabaseName())
```

```
                    .getCollection(mongoConfiguration.
  getCollectionName(), VetReview.class);
      }
}
```

The getCollection() method in this concrete class will get the database and collection from MongoDB. Saving to and deleting from this collection is straightforward, as shown previously. To find it based on a certain condition, we can use find() with eq(). eq() is a BSON filter defined in the MongoDB driver itself.

# Creating a service for the entity

A service will encapsulate any business logic as well as downstream access to the vet-reviews collection (in MongoDB). We can declare an interface to abstract out core methods in a service. To contain a service, we should first create a package called com. packtpub.micronaut.service under the root package.

The VetReviewService interface can then be declared as follows:

```
public interface VetReviewService {

    List<VetReview> findAll();

    VetReview findByReviewId(String reviewId);

    VetReview save(VetReview vetReview);

    void delete(String reviewId);
}
```

The VetReviewService interface will declare barebones methods supported by this service. To provide concrete implementation for these methods, we will create a VetServiceImpl class under the com.packtpub.micronaut.service.impl package:

```
@Singleton
public class VetReviewServiceImpl implements VetReviewService {

    private final VetReviewRepository vetReviewRepository;
```

```java
public VetReviewServiceImpl(VetReviewRepository
    vetReviewRepository) {
        this.vetReviewRepository = vetReviewRepository;
}


@Override
public List<VetReview> findAll() {
        return vetReviewRepository.findAll();
}


@Override
public VetReview findByReviewId(String reviewId) {
        return
          vetReviewRepository.findByReviewId(reviewId);
}


@Override
public VetReview save(VetReview vetReview) {
        return vetReviewRepository.save(vetReview);
}


@Override
public void delete(String reviewId) {
        vetReviewRepository.delete(reviewId);
}
}
```

Concrete implementations for these methods are essentially delegating the call to VetReviewRepository. However, for any future needs, these methods can be extended to contain any business logic, such as data validations or data transformation.

In the next section, we will focus our attention on creating a small command-line utility to perform common CRUD operations on the pet-clinic-reviews collection.

# Performing basic CRUD operations

In order to perform basic CRUD operations on the VetReview entity, we can create a simple utility. We can create a new package called com.packtpub.micronaut. utils to define PetClinicReviewCliClient:

```
@Singleton
@Requires(beans = MongoClient.class)
public class PetClinicReviewCliClient {

    private final Logger log = LoggerFactory.getLogger
      (PetClinicReviewCliClient.class);

    private final VetReviewService vetReviewService;

    public PetClinicReviewCliClient(VetReviewService
      vetReviewService) {
        this.vetReviewService = vetReviewService;
    }
}
```

This utility will inject VetReviewService using the constructor. We will delegate the CRUD calls via this service.

## Performing read/fetch operations

We can define a simple utility method in PetClinicReviewCliClient that can call VetReviewService to fetch all vet reviews or a specific vet review:

```
protected void performFindAll() {
    List<VetReview> vetReviews =
      this.vetReviewService.findAll();
    … iterate over vetReviews
}

protected void performFindByReviewId(String reviewId) {
    VetReview vetReview =
      vetReviewService.findByReviewId(reviewId);
    log.info("Review: {}", vetReview);
}
```

`performFindAll()` will fetch all vet reviews in the collection, whereas `performFindByReviewId()` will fetch a specific review by `reviewId`.

## Performing a save operation

We will save a vet review through a method in `PetClinicReviewCliClient`:

```
protected VetReview performSave() {
    VetReview vetReview = new VetReview(
            UUID.randomUUID().toString(),
            1L,
            3.5,
            "Good experience",
            LocalDate.now());
    return vetReviewService.save(vetReview);
}
```

Though we are defining a vet review object in Java, the constructor in the `VetReview` entity will take care of mapping this object to `BsonDocument` and downstream code in `VetReviewRepository` will save this document to the `vet-reviews` collection.

## Performing a delete operation

In `PetClinicReviewCliClient`, we will define a delete method to `delete` a vet review:

```
protected void performDelete(VetReview vetReview) {
    vetReviewService.delete(vetReview.getReviewId());
}
```

It is a small method and it delegates the call to `VetReviewService` for deletion. `VetReviewService` further invokes the repository to delete the given vet review from the collection.

# Wrapping up

In `Application.java`, we can call `PetClinicReviewCliClient` as a startup event, which can then exhibit basic CRUD operations on the `vet-reviews` collection:

```
@Singleton
public class Application {
```

```
    private final PetClinicReviewCliClient
     petClinicReviewCliClient;

    public Application(PetClinicReviewCliClient
     petClinicReviewCliClient) {
        this.petClinicReviewCliClient =
        petClinicReviewCliClient;
    }

    public static void main(String[] args) {
        Micronaut.run(Application.class, args);
    }

    @EventListener
    void init(StartupEvent event) {
        petClinicReviewCliClient.
         performDatabaseOperations();
    }
 }
```

When we kickstart our application, the aforementioned @EventListener will invoke PetClinicReviewCliClient to perform database operations.

In this section, we covered how to integrate a Micronaut application with MongoDB. We defined the entity, repository, and service to access the MongoDB database. Lastly, we exhibited CRUD operations through a light utility.

# Summary

In this chapter, we covered various aspects of integrating a Micronaut application with relational as well as NoSQL databases. We explored different ways of persistence integration, that is, using ORM (Hibernate), a persistence framework (MyBatis), or a driver-based framework (MongoDB Sync). In each technique, we covered, in depth, how to define entities, relationships, repositories, and services. Each microservice defined a simple command-line utility to exhibit common CRUD operations.

This chapter has given us the skills to cover almost all the data access aspects in Micronaut. In the rest of the book, we will further explore and use these skills and learnings by doing more hands-on exercises and covering other aspects of the Micronaut framework.

In the next chapter, we will work on the web layer of the `pet-clinic` application, defining various REST endpoints in all the microservices.

# Questions

1.  What is an **object-relational mapping (ORM)** framework?

2.  What is Hibernate?

3.  How can you define an entity using Hibernate in Micronaut?

4.  How can you map a one-to-one relationship using Hibernate in Micronaut?

5.  How can you map a one-to-many or many-to-one relationship using Hibernate in Micronaut?

6.  How can you map many-to-many relationships using Hibernate in Micronaut?

7.  How can you perform CRUD operations using Hibernate in Micronaut?

8.  How can you integrate with a relational database using MyBatis in Micronaut?

9.  How can you define MyBatis mappers in Micronaut?

10. How can you perform CRUD operations using MyBatis in Micronaut?

11. How can you integrate with a NoSQL (MongoDB) database in Micronaut?

12. How can you perform CRUD operations in MongoDB in Micronaut?

# 3
# Working on RESTful Web Services

In any microservice development, one of the core aspects is how the microservice interfaces with the external world. **RESTful** or *restful* has emerged as the golden standard of building these service interfaces. RESTful treats all information exchange as a resource exchange among systems. A resource represents the state of the object at the time of the transfer, hence the term **Representational State Transfer** (**REST**). In this chapter, we will explore the key aspects of working on these restful interfaces in the Micronaut framework. We will continue with the controller-service-repository pattern and add restful endpoints to the microservice projects within the pet-clinic application. For outgoing and incoming payloads to these endpoints, we will use **data transfer objects** (**DTOs**) in tandem with MapStruct to map DTOs to/from entities. For hands-on work, we will work to add restful endpoints to the following microservices:

- pet-owner: Working hands-on to add HTTP GET, POST, PUT, and DELETE endpoints for pet-owner schema objects

- pet-clinic: Working hands-on to add HTTP GET, POST, PUT, and DELETE endpoints for pet-clinic schema objects

- pet-clinic-review: Working hands-on to add HTTP GET, POST, PUT, and DELETE endpoints for the pet-clinic-reviews collection

While working on the aforementioned microservices, we will dive into the following topics in this chapter:

- Working on restful microservices in the Micronaut framework
- Using DTOs for the endpoint payloads
- Creating restful endpoints for a microservice
- Using Micronaut's HTTP server APIs
- Using Micronaut's HTTP client APIs

By the end of this chapter, you will have the practical know-how to work on restful web services in the Micronaut framework. This knowledge is important to work on the web layer of a microservice. Furthermore, we will also explore leveraging HTTP server objects and client objects in the Micronaut framework.

# Technical requirements

All the commands and technical instructions in this chapter are run on Windows 10 and macOS. Code examples covered in this chapter are available in the book's GitHub repository at `https://github.com/PacktPublishing/Building-Microservices-with-Micronaut/tree/master/Chapter03`.

The following tools need to be installed and set up in the development environment:

- **Java SDK**: Version 8 or above (we used Java 14).
- **Maven**: This is optional and only required if you would like to use Maven as the build system. However, we recommend having Maven set up on any development machine. Instructions to download and install Maven can be found at `https://maven.apache.org/download.cgi`.
- **Development IDE**: Based on your preferences, any Java-based IDE can be used, but for the purpose of writing this chapter, IntelliJ was used.
- **Git**: Instructions to download and install Git can be found at `https://git-scm.com/downloads`.
- **PostgreSQL**: Instructions to download and install PostgreSQL can be found at `https://www.postgresql.org/download/`.

- **MongoDB**: MongoDB Atlas provides a free online database-as-a-service with up to 512 MB storage. However, if a local database is preferred, then instructions on how to download and install can be found at `https://docs.mongodb.com/manual/administration/install-community/`. We used a local installation for writing this chapter.

- **Rest client**: Any HTTP REST client can be used. We used the Advanced REST Client Chrome plugin.

# Working on restful microservices in the Micronaut framework

In order to learn about restful microservices in the Micronaut framework, we will continue working on the `pet-clinic` application. The following table summarizes the changes we will be making on each of the microservices in the `pet-clinic` application:

| Microservice name | RESTFUL endpoints supported | Database type |
|---|---|---|
| pet-owner | HTTP get, post, put, and delete | Relational (PostgreSQL) |
| pet-clinic | HTTP get, post, put, and delete | Relational (PostgreSQL) |
| pet-clinic-reviews | HTTP get, post, put, and delete | MongoDB |

Table 3.1 – Microservices in the pet-clinic application

In each microservice, we will get hands-on with adding HTTP endpoints for performing CRUD operations on the data objects owned by them.

In our hands-on discussions, we will focus on the following:

- **DTOs**: How DTOs can be used to encapsulate outgoing and incoming payloads for restful endpoints

- **Services**: How services liaison with database repositories for any controller requests

- **Controllers**: How controllers provide a standard restful interface for the external world in the Micronaut framework

Adding to *Figure 2.2* of *Chapter 2, Working on the Data Access*, the following are the components in this chapter within each microservice:

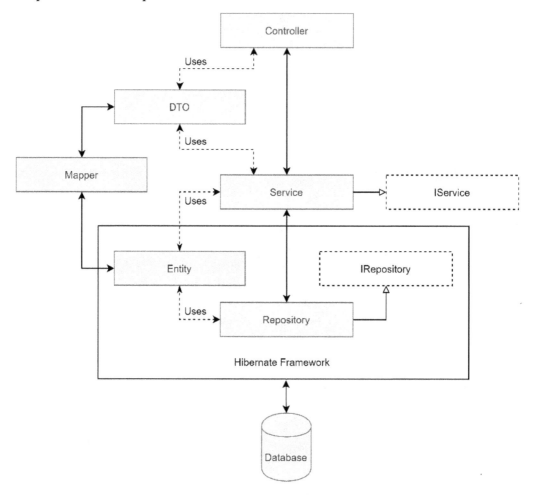

Figure 3.1 – Microservice components

We will continue with the separation of concerns by following the controller-service-repository pattern. For communications between the service and controller, we will explore DTOs. We will work in a bottom-up fashion while covering DTOs, services, and finally controllers. DTOs, mappers, services, and controllers follow the same approach, and therefore, to keep our discussion focused, we will target the pet-owner microservice.

In the next section, our focal point will be DTOs.

# Using DTOs for the endpoint payloads

The DTO pattern originates from the enterprise application architecture and fundamentally, data objects aggregate and encapsulate the data to transfer. Since in the microservices architecture an end client may need varied data from different persistence resources (such as invoice data along with user data), the DTO pattern is very effective in limiting calls to microservices for getting the desired projection of data. DTOs are also known as assembler objects since they assemble data from multiple entity classes.

In this section, we will explore how to implement and map (to an entity) DTOs. In the later sections, we will further dive into using DTOs as an effective mechanism to transfer data from and into a microservice. We will also look at how DTOs can help reduce the number of calls to microservices by assembling data.

## Implementing DTOs

To implement a DTO, we will begin defining a DTO class for the pet owner.

Open the pet-owner microservice project (created in *Chapter 2, Working on the Data Access*) in your preferred IDE. Add the com.packtpub.micronaut.service.dto package to contain all DTOs. We can define OwnerDTO as follows:

```
@Introspected
public class OwnerDTO implements Serializable {
    private Long id;
    private String firstName;
    private String lastName;
    private String address;
    private String city;
    private String telephone;
    private Set<PetDTO> pets = new HashSet<>();
    … getters and setters
}
```

OwnerDTO is implementing the Serializable marker interface to mark that the DTO is serializable. Furthermore, in continuation of our earlier discussion on the assembler pattern, OwnerDTO will also contain a set of PetDTO instances.

Following the similar POJO model, we can define DTOs for other entities in the pet-owner microservice, such as PetDTO, VisitDTO, and PetTypeDTO in the com.packtpub.micronaut.service.dto package.

In the next section, we will work on mapping these DTOs to database entities.

# Using MapStruct to define mappers

**MapStruct** is a code generator that uses annotation processing to implement mappings between source and target Java classes. The implemented MapStruct code consists of plain method calls, therefore it's type-safe and easy to read code. Since we don't need to write code for these mappings, MapStruct is very effective in reducing the source code footprint.

To map DTOs to entities and vice versa, we will use MapStruct in our `pet-owner` microservice. Since we are using Maven, we will have to add the following to the `pom.xml` project:

```
...
<properties>
    <org.mapstruct.version>1.3.1.Final</org.mapstruct.version>
</properties>
...
<dependencies>
    <dependency>
        <groupId>org.mapstruct</groupId>
        <artifactId>mapstruct</artifactId>
        <version>${org.mapstruct.version}</version>
    </dependency>
</dependencies>
```

By importing MapStruct into the project, POM will allow us to leverage the MapStruct toolkit. Furthermore, for the Maven compiler, we will need to add MapStruct to `annotationProcessorPaths`:

```
...
<build>
    <plugins>
        <plugin>
            <groupId>org.apache.maven.plugins</groupId>
            <artifactId>maven-compiler-plugin</artifactId>
            <configuration>
                ...
```

```
                    <annotationProcessorPaths>
                        <path>
                            <groupId>org.mapstruct</groupId>
                            <artifactId>mapstruct-processor</
artifactId>
                            <version>${org.mapstruct.version}</
version>
                        </path>
                        <!-- other annotation processors -->
                    </annotationProcessorPaths>
            <compilerArgs>
                            <arg>-Amapstruct.
defaultComponentModel=jsr330</arg>
                        ....
                    </compilerArgs>
                </configuration>
            </plugin>
        </plugins>
    </build>
```

The annotation processing settings in POM will direct the Java annotation processor to generate source code for any mappings that are marked using MapStruct annotations. In addition, jsr330 is specified as a default component model in the context of the Micronaut framework (in Spring, a Spring model is often used).

We will create a new package named com.packtpub.micronaut.service.mapper to contain all the mapper interfaces. To abstract out a generic entity mapper, we can declare the following interface:

```
public interface EntityMapper <D, E> {
    E toEntity(D dto);
    D toDto(E entity);
    List <E> toEntity(List<D> dtoList);
    List <D> toDto(List<E> entityList);
}
```

The `EntityMapper` interface abstracts out object-to-object and list-to-list conversion methods. By extending this interface, we can easily define an interface to map `OwnerDTO` to the `Owner` entity:

```
@Mapper(componentModel = "jsr330", uses = {PetMapper.class})
public interface OwnerMapper extends EntityMapper<OwnerDTO,
Owner> {
    default Owner fromId(Long id) {
        if (id == null) {
            return null;
        }
        Owner owner = new Owner();
        owner.setId(id);
        return owner;
    }
}
```

The `OwnerMapper` interface extends `EntityMapper` and uses `PetMapper`. `PetMapper` is used to map a set of `PetDTO` instances to `Pet` entities. `PetMapper` can be defined using a very similar approach.

Adhering to the same approach, we can thus define `PetMapper`, `VisitMapper`, and `PetTypeMapper` for the `Pet`, `Visit`, and `PetType` entities, respectively.

So far, we have dived into DTOs and their mappings to corresponding entity classes. In the next section, we will zero down on the service changes concerning DTOs.

## Modifying the services to use DTOs

In *Chapter 2, Working on the Data Access*, to simplify the discussion, we defined service methods that used entity classes directly. This approach is not recommended for a variety of reasons, including the fact that we end up colluding/coupling business logic with database entities or it can become hairy if a service method needs to use multiple entities. Put simply, we must separate the concerns by decoupling database entities from business-required data objects.

To use DTOs in business services, we will need to modify abstracting interfaces. The `OwnerService` interface can be modified to use DTOs as follows:

```
public interface OwnerService {
    OwnerDTO save(OwnerDTO ownerDTO);
    Page<OwnerDTO> findAll(Pageable pageable);
```

```
    Optional<OwnerDTO> findOne(Long id);
    void delete(Long id);
}
```

The save() method is modified to save OwnerDTO instead of the Owner entity. Also, findAll() and findOne() will return OwnerDTO instead of Owner.

Complying with the same approach, we can modify these service interfaces for other entities in the pet-owner microservice, that is, PetService, PetTypeService, and VisitService.

Since we modified abstracting service interfaces, we will have to modify implementing classes too. OwnerServiceImpl can be modified to use DTOs like so:

```
@Singleton
@Transactional
public class OwnerServiceImpl implements OwnerService {
    … injections for OwnerRepository and OwnerMapper
    @Override
    public OwnerDTO save(OwnerDTO ownerDTO) {
        Owner owner = ownerMapper.toEntity(ownerDTO);
        owner = ownerRepository.mergeAndSave(owner);
        return ownerMapper.toDto(owner);
    }
    @Override
    @ReadOnly
    @Transactional
    public Page<OwnerDTO> findAll(Pageable pageable) {
        return ownerRepository.findAll(pageable)
                .map(ownerMapper::toDto);
    }
    @Override
    @ReadOnly
    @Transactional
    public Optional<OwnerDTO> findOne(Long id) {
        return ownerRepository.findById(id)
                .map(ownerMapper::toDto);
    }
    @Override
```

```
    public void delete(Long id) {
        ownerRepository.deleteById(id);
    }
}
```

The save() method will use OwnerMapper to convert OwnerDTO to the entity before the repository method is called. Similarly, fetch methods will convert Owner entities back to OwnerDTOs before returning the responses.

Following the same approach, we can modify PetServiceImpl, PetTypeServiceImpl, and VisitServiceImpl.

So far, we have focused on DTOs and how we can easily use DTOs by mapping them on entity objects using the MapStruct framework. In the next section, we will center our attention on how to create restful endpoints for the pet-owner microservice.

# Creating the restful endpoints for a microservice

Using the Micronaut framework, we can create all commonly used HTTP methods, namely GET, PUT, POST, and DELETE. At the core, all HTTP concerns are encapsulated in the io.micronaut.http package. This package contains HttpRequest and HttpResponse interfaces. These interfaces are the bedrock of defining any HTTP endpoint. Using these standard implementations, we will cover all HTTP methods in the following sections.

## Creating an endpoint for retrieving a list of resources

To create an endpoint for fetching a list of resources, we can begin by adding the com.packtpub.micronaut.web.rest package to contain all controller resources. We will add a fetch all owners endpoint to the OwnerResource controller in the pet-owner microservice:

```
@Controller("/api")
public class OwnerResource {

    ...

    @Get("/owners")
    @ExecuteOn(TaskExecutors.IO)
    public HttpResponse<List<OwnerDTO>>
getAllOwners(HttpRequest request, Pageable pageable) {
        log.debug("REST request to get a page of Owners");
```

```
            Page<OwnerDTO> page = ownerService.findAll(pageable);
            return HttpResponse.ok(page.getContent()).
headers(headers ->
            PaginationUtil.
generatePaginationHttpHeaders(headers, UriBuilder.of(request.
getPath()), page));
    }
    ...
}
```

There are a few things we need to ponder here:

1. @Controller: This is a stereotype that marks OwnerResource as a restful
   controller and is exposed on the /api base URL.

2. @Get: This annotation is used to expose the getAllOwners() method on
   HTTP get over the relative URL of /owners.

3. @ExecuteOn: This annotation specifies to execute this GET request on the I/O
   thread pool. The value of the @ExecuteOn annotation can be any executor
   defined in micronaut.executors.

4. HttpRequest: getAllOwners() takes on HttpRequest as an input
   parameter. Often, HttpRequest is passed to trace and log details of the origin,
   for example, which IP address is submitting the request.

5. pageable: This is a standard interface in the io.micronaut.data.model
   package and is often used to pass pagination information in the request, for
   example, .../api/owners?page=1&size=50&sort=(firstName). This
   pagination information is often optional and passed as query parameters.

6. HttpResponse: getAllOwners() returns HttpResponse, which is a generic
   interface and embodies a specific type of response within the example. In this
   example, it is returning List<OwnerDTO>.

7. PaginationUtil: This is a custom class in the com.packtpub.micronaut.
   util package and it generates paged responses.

We can run the pet-owner service locally and by default, it will run on port 8080. We can use any REST client and hit http://localhost:8080/api/owners and we will get the following response:

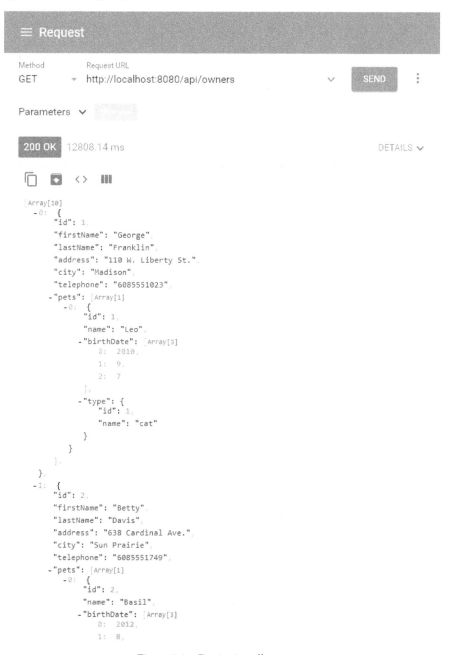

Figure 3.2 – Retrieving all owners

As shown in the preceding screenshot, when we call the fetch all owners endpoint, we are not passing any pagination information to get all owners. Therefore, it will retrieve all the owners and their pet information from the database.

## Creating an endpoint for retrieving a specific resource

To create an endpoint for fetching a specific resource, we will add a method in OwnerResource to fetch a specific owner, as follows:

```
@Get("/owners/{id}")
@ExecuteOn(TaskExecutors.IO)
public HttpResponse<Optional<OwnerDTO>> getOwner(@PathVariable
Long id) {
    log.debug("REST request to get Owner : {}", id);
    return ownerService.findOne(id);
}
```

There are a few things we need to ponder here:

1. @PathVariable: This is used to specify and match a path variable in the HTTP call. In the proceeding example, we are passing the owner ID as a path variable.

2. Optional: In this example, we are returning an optional object. If the service can find an owner for the specified ID, then an HTTP 200 response will be sent; otherwise, HTTP 404 will be reverted.

We can run the `pet-owner` service locally and by default, it will run on port `8080`. We can use any rest client and by hitting `http://localhost:8080/api/owners/1`, we will get the following response:

Figure 3.3 – Retrieving a specific owner

`OwnerDTO` encapsulated the pet data as well so the response payload will fetch complete details about the owner with ID `1`. Furthermore, if any non-existing ID is passed to the proceeding HTTP request, then it will result in `HTTP 404`:

Figure 3.4 – Retrieving a non-existing owner

In the preceding API call, we tried to fetch an owner with the `123789` ID, but since we don't have an owner with this ID, it results in an `HTTP 404 not found` response.

## Creating an endpoint for inserting a resource

To create an endpoint to insert a resource, we will add an HTTP POST method to `OwnerResource` as follows:

```
...
@Post("/owners")
@ExecuteOn(TaskExecutors.IO)
public HttpResponse<OwnerDTO> createOwner(@Body OwnerDTO
ownerDTO) throws URISyntaxException {
    if (ownerDTO.getId() != null) {
```

```
        throw new BadRequestAlertException("A new owner cannot
already have an ID", ENTITY_NAME, "idexists");
    }
    OwnerDTO result = ownerService.save(ownerDTO);
    URI location = new URI("/api/owners/" + result.getId());
    return HttpResponse.created(result).headers(headers -> {
        headers.location(location);
        HeaderUtil.createEntityCreationAlert(headers,
applicationName, true, ENTITY_NAME, result.getId().toString());
    });
}
...
```

There are a few things to pay attention to here:

1. @Post: createOwner() is exposed using the @Post annotation as an
   HTTP POST API.

2. @ExecuteOn: This annotation specifies to execute this POST request on the I/O
   thread pool.

3. @Body: The @Body annotation specifies that the ownerDTO method argument in
   createOwner() is bound to the body of the incoming HTTP POST request.

4. Not null id check: Using an if construct, we are quickly checking that the
   HTTP body doesn't contain a payload with the ID already defined. This is a case
   of asserting a business validation and if it fails, then the API is throwing a bad
   request exception.

5. HttpResponse.created: In the happy path scenario, the API will return HTTP
   201 created. This response indicates that the request was executed successfully and
   a resource was created.

We can boot up the pet-owner microservice locally and hit an HTTP POST request in
a rest client as follows:

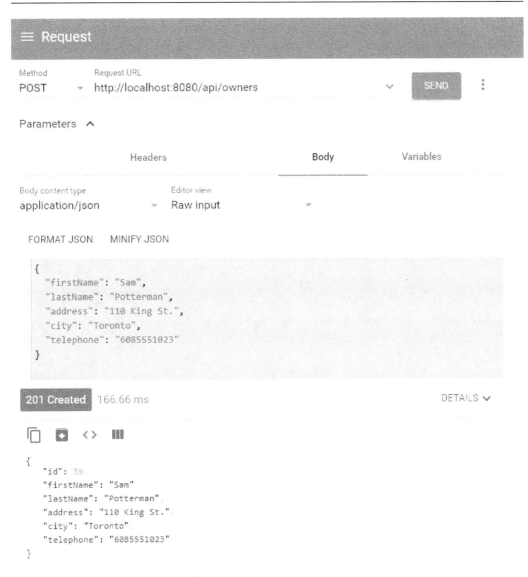

Figure 3.5 – Inserting an owner

In the preceding HTTP POST call, we passed an owner object to be inserted in the HTTP body. As anticipated, the API call was successful and returned an HTTP 201 created response.

Furthermore, if we try to hit this HTTP POST method to create an owner that has an ID value already, then it will fail:

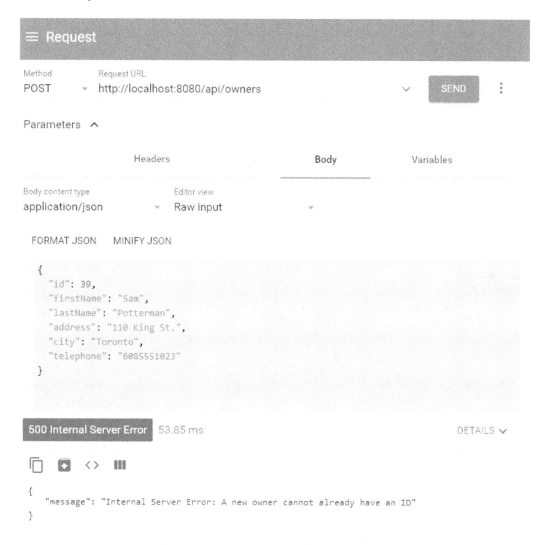

Figure 3.6 – Inserting an existing owner again

When we try to insert an owner with an ID value in the body payload, then an HTTP 500 internal server error is thrown with a message – Internal Server Error: A new owner cannot already have an ID.

# Creating an endpoint for updating a resource

To create an endpoint to update a resource, we will add an HTTP PUT method to `OwnerResource` as follows:

```
...
@Put("/owners")
@ExecuteOn(TaskExecutors.IO)
public HttpResponse<OwnerDTO> updateOwner(@Body OwnerDTO
ownerDTO) throws URISyntaxException {
    log.debug("REST request to update Owner : {}", ownerDTO);
    if (ownerDTO.getId() == null) {
        throw new BadRequestAlertException("Invalid id",
ENTITY_NAME, "idnull");
    }
    OwnerDTO result = ownerService.save(ownerDTO);
    return HttpResponse.ok(result).headers(headers ->
            HeaderUtil.createEntityUpdateAlert(headers,
applicationName, true, ENTITY_NAME, ownerDTO.getId().
toString()));
}
...
```

A few things to ponder here:

1.  `@Put`: The `@Put` annotation exposes the `updateOwner()` method as an HTTP PUT API.

2.  `@ExecuteOn`: This annotation specifies to execute this PUT request on the I/O thread pool. The I/O thread pool is a mechanism to achieve I/O concurrency by maintaining a quorum of threads waiting to be assigned the I/O requests.

3.  `@Body`: The `@Body` annotation specifies that the `ownerDTO` method argument in `createOwner()` is bound to the body of the incoming HTTP PUT request.

4.  `Null id check`: Using an `if` construct, we are quickly checking that the HTTP body doesn't contain a payload with a null ID. If the ID is null, then the API is throwing a bad request exception.

5.  `HttpResponse.ok`: In the happy path scenario, the API will return `HTTP 200`. This response indicates that the request was executed successfully and a resource was updated.

We can boot up the `pet-owner` microservice locally and hit an HTTP PUT request in a rest client like so:

Figure 3.7 – Updating an owner

In the HTTP PUT request, we requested to update a resource we just inserted previously. We are updating the address and city for this resource. As anticipated, the request was executed successfully and an HTTP 200 response was returned.

If we try to update a resource with a null ID, then it will fail:

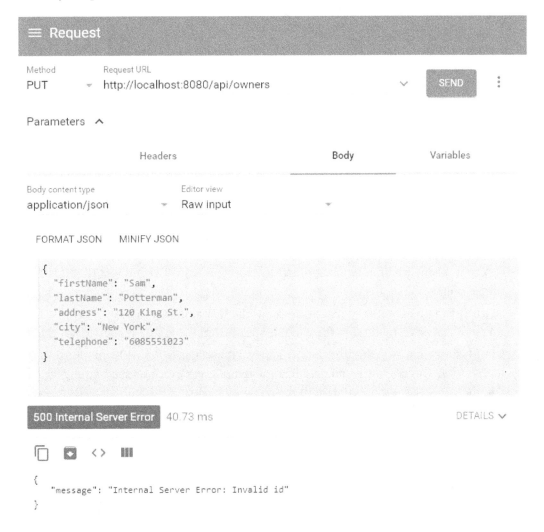

Figure 3.8 – Updating a non-existing owner

In the preceding HTTP PUT call, we tried to update an owner with a null ID. An HTTP 500 internal server error is thrown with a message – Internal Server Error: Invalid id.

# Creating an endpoint for deleting a resource

To exhibit deleting a resource, we can add an HTTP DELETE method to
`OwnerResource` as follows:

```
...
@Delete("/owners/{id}")
@ExecuteOn(TaskExecutors.IO)
public HttpResponse deleteOwner(@PathVariable Long id) {
    log.debug("REST request to delete Owner : {}", id);
    ownerService.delete(id);
    return HttpResponse.noContent().headers(headers ->
HeaderUtil.createEntityDeletionAlert(headers, applicationName,
true, ENTITY_NAME, id.toString())));
}
...
```

A few things to ponder here:

1.  `@Delete`: The `@Delete` annotation exposes the `deleteOwner()` method as an
    HTTP DELETE API.

2.  `@ExecuteOn`: This annotation specifies to execute this PUT request on the I/O
    thread pool.

3.  `@PathVariable`: `@PathVariable` binds the ID parameter in the
    `deleteOwner()` method to the variable specified in the HTTP request URL.

4.  `HttpResponse.noContent`: In the happy path scenario, the API will return
    `HTTP 204`. This response indicates that the request was executed successfully
    and there is no additional content in the response payload.

We can boot up the `pet-owner` microservice locally and hit an HTTP DELETE request
in a rest client:

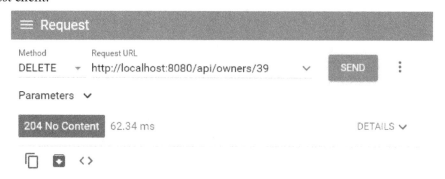

Figure 3.9 – Deleting an owner

We requested to delete the owner that we inserted/updated in previous examples. As anticipated, the request was successful and an HTTP 204 response was returned.

So far, we have explored how to create all different types of restful endpoints. This discussion paves our way into using Micronaut's HTTP server APIs. In the next section, we will cover some practical aspects of leveraging these server APIs.

# Using Micronaut's HTTP server APIs

The Micronaut framework offers a non-blocking HTTP server based on Netty. These server APIs can be leveraged to address some useful microservice requirement scenarios. To do any hands-on work, we will continue with the pet-owner microservice.

At the outset, we must have the following Maven dependency in the project POM:

```
...
<dependency>
<groupId>io.micronaut</groupId>
<artifactId>micronaut-http-server-netty</artifactId>
<scope>compile</scope>
</dependency>
...
```

This dependency should already be in the pet-owner project. In the next sections, we will explore some useful configurations in the HTTP server world.

## Binding HTTP requests in the Micronaut framework

There are several ways we can bind arguments to an HTTP request in the Micronaut framework:

- @Body: As discussed before, @Body binds the argument from the body of the HTTP request.

- @CookieValue: This binds an argument from the cookie value in the HTTP request.

- @Header: This binds an argument from a header in the HTTP request.

- @QueryValue: This binds an argument from a query parameter in the HTTP request.

- @PathVariable: This binds an argument from the path in the HTTP request.

- @Part: In a multi-part HTTP request, this binds the argument to a part.

- @RequestBean: This binds any object (in the HTTP request) to a Java bean.

Along with the preceding HTTP request bindings, the Micronaut framework supports multiple URI templates. These templates come in handy in creating various kinds of restful microservices. Considering the owner resource, we can support the following URI template matching in Micronaut:

| URI template | Description | Example |
|---|---|---|
| /owners/{ID} | Fetch owner resource with a matching ID. | /owners/1 |
| /owners/{ID:4} | Fetch owner resource with a matching ID, where the ID can assume a maximum of four digits. | /owners/321 |
| /owners{/ID} | Fetch owners, if ID (optional) is provided then match an owner with the specified ID. | /owners or /owners/1 |
| /owners{?pageNumber, pageSize,locale} | Fetch owners, if optional query parameters are specified then filter and serve the request based on query parameters. | /owners?pageNumber =10,pageSize=10,locale=fr |
| /owners/{ID:regex} ... /owners/{ID:[0-9]+} | Fetch owner resource with a matching ID, where the ID can assume a value only according to the specified regex. | /owners/1 |

Table 3.2 – URI templates in the Micronaut framework

In the preceding table, we can see various approaches in matching/restricting an incoming HTTP request and what each approach can achieve for a specific resource. In the next section, we will discuss how we can validate the data in an HTTP request.

# Validating data

The Micronaut framework supports both JSR 380 bean validations as well as hibernate bean validations. By default, a Micronaut project usually contains a micronaut-validation dependency, which is implemented based on the JSR 380 standard.

Using validation annotations such as @NotNull and @NotEmpty, we can verify whether an argument/parameter is meeting all the validation criteria or not before processing a request. Some useful annotations are as follows:

- @NotNull: This asserts if the argument/parameter value is not null, for example, @NotNull @PathVariable String ownerId.

- @NotEmpty: This asserts if the argument is not null and not empty. It can be applied to any argument of the String, Collection, or Map type, and so on, for example, @NotEmpty @PathVariable String ownerId.

- `@Min`: This validates that the annotated property has a value greater than or equal to the `value` attribute, for example, `@Min(value = 0) Integer offset`.

- `@Max`: This validates that the annotated property has a value equal to or smaller than the `value` attribute, for example, `@Max(value =100) Integer offset`.

- `@Valid`: `@Valid` validates all arguments/parameters within the object graph. It recursively scans all inner `@Valid` usages within the object graph and determines the final validation after checking everything.

If any of these validations are not met, then `javax.validation.ConstraintViolationException` is thrown.

To handle constraint violation exceptions as well as other checked/unchecked exceptions, we will explore some options in the next section.

## Handling errors

The Micronaut framework provides good coverage on raising different kinds of exceptions throughout the life cycle of an HTTP request along with handling these exceptions. The standard exception handlers can be found in the `io.micronaut.http.server.exceptions` package. By default, the following handlers are provided within this package:

- `ContentLengthExceededHandler`: This handles `ContentLengthExceededException` by returning an `HttpStatus.REQUEST_ENTITY_TOO_LARGE` response.

- `ConversionErrorHandler`: This handles `ConversionErrorException` by returning an `HttpStatus.BAD_REQUEST` response.

- `HttpStatusHandler`: This handles `HttpStatusException` by returning `HttpStatus` as specified in the `HttpStatusException` response.

- `JsonExceptionHandler`: This handles `JsonProcessingException` by returning an `HttpStatus.BAD_REQUEST` response.

- `UnsatisfiedArgumentHandler`: This handles `UnsatisfiedArgumentException` by returning an `HttpStatus.BAD_REQUEST` response.

- `URISyntaxHandler`: This handles `URISyntaxException` by returning an `HttpStatus.BAD_REQUEST` response.

In addition to the preceding standard exceptions and exception handlers, we can create our custom exceptions and exception handlers.

We can assume a hypothetical `FooException` that can be raised by a microservice:

```
public class FooException extends RuntimeException {
...
}
```

`FooException` extends `RuntimeException`, which makes it an unchecked exception, and we can create `FooExceptionHandler` to handle this exception:

```
@Produces
@Singleton
@Requires(classes = {FooException.class, ExceptionHandler.
class})
public class FooExceptionHandler implements
ExceptionHandler<FooException, HttpResponse> {
    @Override
    public HttpResponse handle(HttpRequest request,
FooException exception) {
        JsonError error = new JsonError(exception.
getMessage());
        error.path('/' + exception.getArgument().getName());
        error.link(Link.SELF, Link.of(request.getUri()));
        return HttpResponse.status(HttpStatus.INTERNAL_SERVER_
ERROR).body(error);
    }
}
```

The `FooExceptionHandler` bean injection mandates the required injection of `FooException` and `ExceptionHandler`. At any place within the code base, if `FooException` is raised, it will be caught by `FooExceptionHandler` and a `HttpStatus.INTERNAL_SERVER_ERROR` response will be returned.

## Versioning the APIs

Micronaut supports API versioning by using the `@Version` annotation. This annotation can be used at the controller or method level.

Versioning is not supported by default and to enable versioning, we must make the following changes in `application.yml`:

```yaml
micronaut:
  application:
    name: Pet-Owner
  router:
    versioning:
      enabled: true
      default-version: 1
....
```

In the configuration, we have enabled versioning and by default, if no version is specified, then the incoming request will be served by the version 1 API.

To exhibit versioning, we will add `@Version` to the `getOwner()` method of `OwnerResource` in the `pet-owner` microservice:

```java
@Version("1")
@Get("/owners/{id}")
@ExecuteOn(TaskExecutors.IO)
public Optional<OwnerDTO> getOwner(@PathVariable Long id) {
    log.debug("REST request to get Owner : {}", id);
    return ownerService.findOne(id);
}
...
@Version("2")
@Get("/owners/{id}")
@ExecuteOn(TaskExecutors.IO)
public Optional<OwnerDTO> getOwnerV2(@PathVariable Long id) {
    log.debug("REST request to get Owner : {}", id);
    return ownerService.findOne(id);
}
```

`@Version` allows multiple projections of the `getOwner()` method. This annotation comes in handy in supporting event-driven microservices.

To test these changes, we can run `pet-owner` locally and use the advanced rest client to make a call to `getOwner()`:

Figure 3.10 – Calling a versioned API

In the preceding HTTP call, we specified the version using the `X-API-VERSION` header. As per the value specified in this header, this call will be served by the version 2 API.

So far, we have explored how to leverage HTTP server APIs in the Micronaut framework. In the next section, we will focus our attention on HTTP client aspects.

# Using Micronaut's HTTP client APIs

Micronaut's HTTP client is a non-blocking client based on Netty with baked-in cloud features such as service discovery and load balancing. This custom implementation enhances the standard HTTP client with microservices architecture.

To exhibit basic HTTP calls, we will create an HTTP client for `OwnerResource` in the `pet-owner` microservice. At the outset, we must have the `micronaut-http-client` dependency in the `pom.xml` project:

```
...
<dependency>
<groupId>io.micronaut</groupId>
<artifactId>micronaut-http-client</artifactId>
<scope>compile</scope>
</dependency>
...
```

`micronaut-http-client` encapsulated all HTTP client-related APIs, such as creating `RxHttpClient`, performing all HTTP operations, and handling and processing response payloads.

Next, we will explore how to leverage `micronaut-http-client` to perform various HTTP operations on the endpoint we created earlier.

## Performing an HTTP PUT operation

To perform an HTTP PUT operation, we can create `OwnerResourceClient`. This client can be packaged inside `com.packtpub.micronaut.web.rest.client`.

We can add the following method to perform an HTTP call:

```
public List<OwnerDTO> getAllOwnersClient() throws
MalformedURLException {
    HttpClient client = HttpClient.create(new URL("https://" +
server.getHost() + ":" + server.getPort()));
    OwnerDTO[] owners = client.toBlocking().
retrieve(HttpRequest.GET("/api/owners"), OwnerDTO[].class);
    return List.of(owners);
}
```

`HttpClient.create()` will create an HTTP client that we then use to make an HTTP GET request.

# Performing an HTTP POST operation

To perform an HTTP POST operation, we can add the following method to
`OwnerResourceClient`:

```
public OwnerDTO createOwnerClient() throws
MalformedURLException {
    HttpClient client = HttpClient.create(new URL("https://" +
server.getHost() + ":" + server.getPort()));

    OwnerDTO newOwner = new OwnerDTO();
    newOwner.setFirstName("Lisa");
    newOwner.setLastName("Ray");
    newOwner.setAddress("100 Queen St.");
    newOwner.setCity("Toronto");
    newOwner.setTelephone("1234567890");

    return client.toBlocking().retrieve(HttpRequest.POST("/api/
owners", newOwner), OwnerDTO.class);
}
```

`HttpClient.create()` will create an HTTP client that we then use to make an HTTP
POST request. An owner object will be passed as the request payload.

# Performing an HTTP PUT operation

To perform an HTTP PUT operation, we can add the following method to
`OwnerResourceClient`:

```
public OwnerDTO updateOwnerClient() throws
MalformedURLException {
    HttpClient client = HttpClient.create(new URL("https://" +
server.getHost() + ":" + server.getPort()));

    OwnerDTO owner = new OwnerDTO();
    owner.setId(1L);
    owner.setAddress("120 Queen St.");

    return client.toBlocking().retrieve(HttpRequest.PUT("/api/
owners", owner), OwnerDTO.class);
}
```

`HttpClient.create()` will create an HTTP client that we then use to make an HTTP PUT request. An `owner` object will be passed as the request payload.

## Performing an HTTP DELETE operation

To perform an HTTP DELETE operation, we can add the following method to `OwnerResourceClient`:

```
public Boolean deleteOwnerClient() throws MalformedURLException
{
    HttpClient client = HttpClient.create(new URL("https://" +
server.getHost() + ":" + server.getPort()));
    long ownerId = 1L;
    HttpResponse httpResponse = client.toBlocking().
retrieve(HttpRequest.DELETE("/api/owners" + ownerId),
HttpResponse.class);
    return httpResponse.getStatus().equals(HttpStatus.NO_
CONTENT);
}
```

`HttpClient.create()` will create an HTTP client that we then use to make an HTTP DELETE request. A no content message will be returned if the request executes successfully.

# Summary

In this chapter, we covered various aspects of working on web endpoints in a Micronaut application. We kick-started with the concept of assemblers or DTOs and then dived into how to create restful endpoints for supporting essential CRUD operations. Also, we experimented with some of the HTTP server APIs in the Micronaut framework. Lastly, we focused on the HTTP client aspects and created a client utility using `micronaut-http-client`.

This chapter has given us the skills related to various practical aspects of working on restful microservices in the Micronaut framework. Furthermore, by exploring the HTTP client, we covered these aspects end to end. This hands-on knowledge to work on the web layer is pivotal in developing any microservice.

In the next chapter, we will work on securing the web layer of the `pet-clinic` microservice by experimenting with different approaches and methods in safeguarding the restful endpoints.

# Questions

1.  What is a DTO?
2.  How do you use MapStruct in the Micronaut framework?
3.  How do you create a restful HTTP GET endpoint in the Micronaut framework?
4.  How do you create a restful HTTP POST endpoint in the Micronaut framework?
5.  How do you create a restful HTTP PUT endpoint in the Micronaut framework?
6.  How do you create a restful HTTP DELETE endpoint in the Micronaut framework?
7.  How does Micronaut support HTTP request binding?
8.  How can we validate data in the Micronaut framework?
9.  How can we version the APIs in the Micronaut framework?
10. How does Micronaut support HTTP client aspects?
11. How can we create an HTTP client in the Micronaut framework?

# 4
# Securing the Microservices

Protecting microservices' interfaces as well as the world encompassed by them is a crucial facet for any application development. Various topologies, tools, and frameworks have arisen in recent times to address the security aspects of web services/microservices. In the course of this chapter, we will dive into some core and often-used security paradigms in microservices. We will continue with the `pet-clinic` application from the previous chapter. For hands-on work, we will work toward securing microservices while covering the following authentication strategies in the Micronaut framework:

- `pet-owner`: Working hands-on to secure `pet-owner` microservice endpoints using **session authentication**

- `pet-clinic`: Working hands-on to secure `pet-clinic` microservice endpoints using **JWT authentication**

- `pet-clinic-review`: Working hands-on to secure `pet-clinic-review` microservice endpoints using **OAuth authentication**

With the aforementioned hands-on exercises, we will be covering the following topics in this chapter:

- Using **session authentication** to secure the service endpoints

- Using **JWT authentication** to secure the service endpoints

- Using **OAuth** to secure the service endpoints

By the end of this chapter, you will have a nifty knowledge of working with various authentication strategies and local or cloud identity providers in the Micronaut framework.

# Technical requirements

All the commands and technical instructions in this chapter are run on Windows 10 and Mac OS X. Code examples covered in this chapter are available in the book's GitHub repository at `https://github.com/PacktPublishing/Building-Microservices-with-Micronaut/tree/master/Chapter04`.

The following tools need to be installed and set up in the development environment:

- **Java SDK**: Version 13 or above (we used Java 14).

- **Maven**: This is optional and only required if you would like to use Maven as the build system. However, we recommend having Maven set up on any development machine. Instructions to download and install Maven can be found at `https://maven.apache.org/download.cgi`.

- **Development IDE**: Based on your preference, any Java-based IDE can be used, but for the purpose of writing this chapter, IntelliJ was used.

- **Git**: Instructions to download and install Git can be found at `https://git-scm.com/downloads`.

- **PostgreSQL**: Instructions to download and install PostgreSQL can be found at `https://www.postgresql.org/download/`.

- **MongoDB**: MongoDB Atlas provides a free online database-as-a-service with up to 512 MB storage. However, if a local database is preferred, then instructions to download and install can be found at `https://docs.mongodb.com/manual/administration/install-community/`. We used a local installation for this chapter.

- **Rest client**: Any HTTP rest client can be used. We used the Advanced REST Client Chrome plugin.

- **Docker**: Instructions to download and install Docker can be found at `https://docs.docker.com/get-docker/`.

- **OpenSSL**: Instructions to download and install OpenSSL can be found at `https://www.openssl.org/source/`.

# Working on RESTful microservices in the Micronaut framework

In order to dive into the security aspects of the Micronaut framework, we will continue working on the `pet-clinic` application. The following table summarizes the changes we will be making to secure each of the microservices in the application:

| Microservice name | Authentication strategy | Identity provider |
|---|---|---|
| pet-owner | Session | Local |
| pet-clinic | JWT | Keycloak |
| pet-clinic-reviews | OAuth | Cloud SaaS identity provider (Okta) |

Table 4.1 – Securing the microservices in the pet-clinic application

To secure the desired endpoints in the microservices, we will focus on the following two key aspects:

- **Identity provider**: Essentially, an identity provider owns the concerns regarding storing and maintaining digital identities. Furthermore, it resolves any security claim by authenticating the submitted digital identity with its quorum of stored identities.

- **Authentication strategy**: The authentication strategy will dictate how a microservice will communicate with the identity provider to authenticate and authorize the user requests.

Adding to the diagram of the components from *Chapter 3, Working on Restful Web Services*, the following will be the changes in this chapter within each microservice:

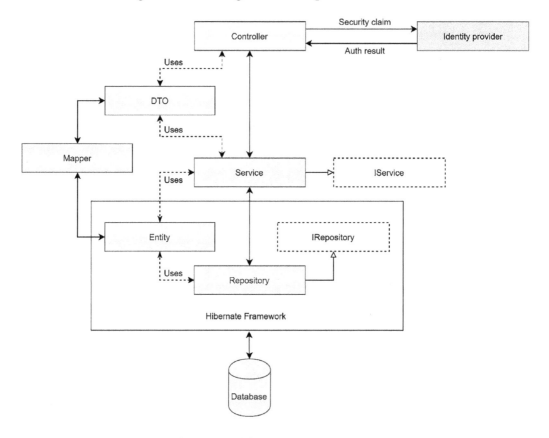

Figure 4.1 – Microservice components

We will stick to our usual pattern of separating the concerns. We will use an identity provider in tandem with an authentication strategy within each of the microservices.

In the next section, our focus will be to cover out-of-the-box tools provided by the Micronaut framework for security concerns.

## The basics of Micronaut security

For handling any security aspects, the Micronaut framework has a built-in `SecurityFilter` object. The `SecurityFilter` object intercepts any incoming HTTP requests and kickstarts the authentication/authorization process as configured in the application. In the following diagram, you can see the workflow within the `SecurityFilter` object for authorizing a user request:

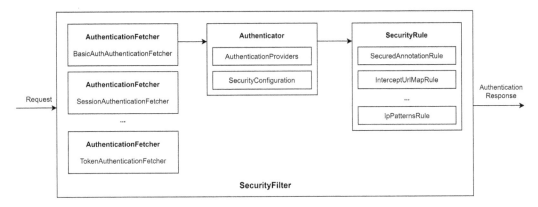

Figure 4.2 – Micronaut security filter

Micronaut's `SecurityFilter` has three essential parts:

- **AuthenticationFetcher**: `AuthenticationFetcher` will fetch the required downstream authenticator for authenticating the user request.

- **Authenticator**: `Authenticator` injects the configured authentication provider(s) and security configurations for authenticating the user request. An `AuthenticationResponse` object is created based on the success or failure of the auth operation.

- **SecurityRule**: If auth is successful, then the security filter will further invoke security rules. An application can configure one or more security rules either using out-of-the-box security rules such as `SecuredAnnotationRule` or `IpPatternsRule` or by creating its own security rules by extending `AbstractSecurityRule`. If the request satisfies all the security rules, then a successful `AuthenticationResponse` response is returned by the security filter; otherwise, it will return a failed `AuthenticationResponse` response.

By leveraging `SecurityFilter`, in the next section, we will focus on how to secure a microservice using session authentication in the Micronaut framework.

# Securing service endpoints using session authentication

In session-based authentication, the user state is stored at the server side. When a user logs in to the server, the server starts the session and issues a session ID in a cookie. The server uses the session ID to uniquely identify a session from the session quorum. Any subsequent user requests must have this session ID passed as a cookie to resume the session:

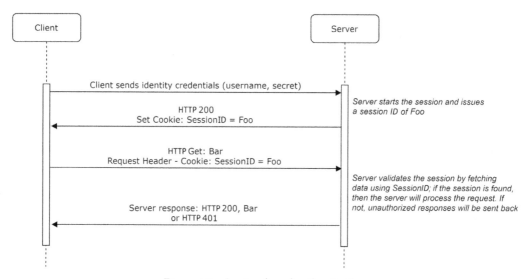

Figure 4.3 – Session-based authentication

As shown in the preceding figure, in a session-based authentication strategy, the server does the heavy lifting of keeping a track of the session. A client must provide a valid session ID to resume the session.

To learn how to secure a microservice using session-based authentication, we will experiment on the pet-owner microservice. To begin, we will need to enable security by adding the following dependencies to the pom.xml project:

```
<!-- Micronaut security -->
    <dependency>
        <groupId>io.micronaut.security</groupId>
        <artifactId>micronaut-security</artifactId>
    </dependency>
    <dependency>
```

```
    <groupId>io.micronaut.security</groupId>
    <artifactId>micronaut-security-session</artifactId>
  </dependency>
...
```

By importing the `micronaut-security` and `micronaut-security-session` dependencies, we can leverage the session authentication toolkit in the `pet-owner` microservice. Once these dependencies are imported, we will then need to configure `application.properties` as shown in the next code block:

```
security:
    enabled: true
    authentication: session
    session:
      enabled: true


    # Auth endpoint
    endpoints:
      login:
        enabled: true
      logout:
        enabled: true
```

As mentioned in the preceding `application.properties` instance, we will enable the security by setting `enabled` to `true` and specifying `session` as the desired authentication strategy. Furthermore, the Micronaut security toolkit provides `LoginController` and `LogoutController` out of the box. In the application properties, we have enabled them and since we haven't specified a custom path for these controllers, they will be accessible at default specified paths of .../`login` and .../`logout`, respectively.

We will use a basic local identity provider that will leverage application properties to store user data. This is very primitive but will help in simplifying learning and exploration. Let's add some user data to `application.properties`:

```
identity-store:
  users:
    alice: alice@1
    bob: bob@2
    charlie: charlie@3
```

```
roles:
  alice: ADMIN
  bob: VIEW
  charlie: VIEW
```

We have added three users: alice, bob, and charlie. Each user is also assigned a role for the pet-owner microservice.

In the next section, we will explore how to implement an authentication provider that will use the configured application properties for user data.

## Implementing a basic authentication provider

To implement a basic authentication provider, we will begin by creating a com.packtpub.micronaut.security security package. This package will encompass all the artifacts concerning security.

We will first add IdentityStore to this package:

```
@ConfigurationProperties("identity-store")
public class IdentityStore {
    @MapFormat
    Map<String, String> users;
    @MapFormat
    Map<String, String> roles;
    public String getUserPassword(String username) {
        return users.get(username);
    }
    public String getUserRole(String username) {
        return roles.get(username);
    }
}
```

The IdentityStore class maps to the application properties for accessing the user data. We can leverage this identity store to implement the authentication provider, as shown in the following code snippet:

```
@Singleton
public class LocalAuthProvider implements
AuthenticationProvider {
    @Inject
    IdentityStore store;
    @Override
    public Publisher<AuthenticationResponse>
      authenticate(HttpRequest httpRequest,
      AuthenticationRequest authenticationRequest) {
        String username =
         authenticationRequest.getIdentity().toString();
        String password =
         authenticationRequest.getSecret().toString();
        if (password.equals(store.getUserPassword
         (username))) {
            UserDetails details = new UserDetails
            (username, Collections.singletonList
            (store.getUserRole(username)));
            return Flowable.just(details);
        } else {
            return Flowable.just(new
            AuthenticationFailed());
        }
    }
}
```

`LocalAuthProvider` implements the standard `AuthenticationProvider` interface by concretely defining the `authenticate()` method. In the `authenticate()` method, we simply check whether the identity and secret specified in the user request match any username and password in the identity store. If we find a match, then we return the `UserDetails` object, else we return `AuthenticatonFailed`.

In the next section, we will concentrate on how we can configure authorizations for the `pet-owner` endpoints.

# Configuring authorizations for the service endpoints

Often in the user requirements for a microservice, there will be scenarios where we need anonymous as well as secured access. To begin with, we will provide anonymous access to `PetResource` and `VisitResource`.

There are two ways to provide anonymous access in Micronaut security:

- Using `@Secured(SecurityRule.IS_ANONYMOUS)`
- Configuring `intercept-url-map` in the application properties

In the following sections, we will drill down into both approaches.

### Granting anonymous access using SecurityRule.IS_ANONYMOUS

Micronaut security has a built-in anonymous access security rule. To give access to the whole controller or limit it to a specific endpoint, we can simply use the `@Secured` annotation. In `PetResource`, we have given anonymous access to all the endpoints by using this annotation at the controller level:

```
@Controller("/api")
@Secured(SecurityRule.IS_ANONYMOUS)
public class PetResource {

....

}
```

Using `@Secured(SecurityRule.IS_ANONYMOUS)` allows anonymous access to all the `PetResource` endpoints. We can simply boot the service and try accessing any `PetResource` endpoint. You can use any REST client to hit the endpoint. In the following screenshot, you'll notice how we are using a rest client to make the HTTP GET call:

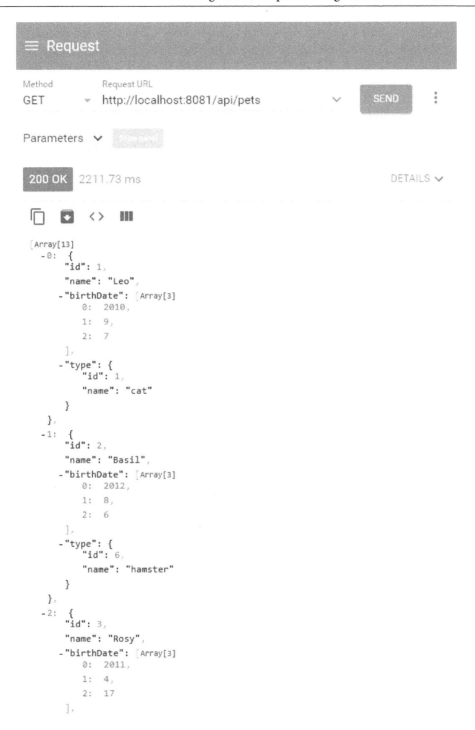

Figure 4.4 – Anonymous access to pets

As shown in the preceding screenshot, we can anonymously access `PetResource` as it is configured for anonymous access using `@Secured(SecurityRule.IS_ANONYMOUS)`.

In the next section, we will see how we can grant anonymous access using application properties.

## Granting anonymous access using application properties

We can also configure anonymous access to a controller or specific endpoint in the controller using application properties. In the following code snippet, we are configuring anonymous access to `.../api/visits` endpoints:

```
# Intercept rules
    intercept-url-map:
      - pattern: /api/visits
        access: isAnonymous()
```

In the application properties, we have configured that any user request to `.../api/visits` should be granted anonymous access. This will allow all users (authenticated as well as unauthenticated) to access `VisitResource`.

To quickly test that we can access `.../api/vistis` anonymously, we can try hitting any `VisitResource` endpoint:

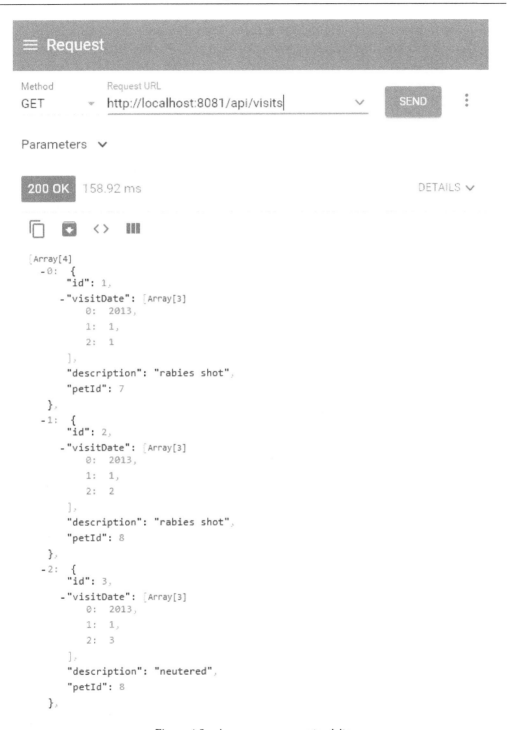

Figure 4.5 – Anonymous access to visits

As shown in the preceding screenshot, we can anonymously access `VisitResource` as it is configured for anonymous access using `intercept-url-map` in `application.properties`.

In the next section, we will explore how to grant secure access using the earlier-defined authentication provider.

## Granting secured access using the local identity provider

To grant secured access, we can use the `@Secured` annotation as well as `intercept-url-map`. In this hands-on `OwnerResource`, we will define secured access to `OwnerResource` using the `@Secured` annotation. Check out the following code block:

```
@Controller("/api")
@Secured(SecurityRule.IS_AUTHENTICATED)
public class OwnerResource {
    ...
}
```

All the endpoints within `OwnerResource` are granted only secured access. If we try to hit any …/`owners` endpoint, the microservice will return a forbidden response, as shown next:

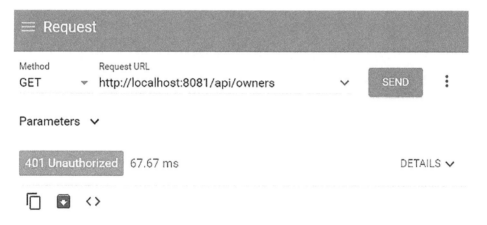

Figure 4.6 – Unauthenticated access to owners

As shown in the preceding screenshot, if we try to access any owner endpoint without specifying identity credentials, the microservice will throw an HTTP 401 Unauthorized response.

For successful access to the owner endpoints, we will need to obtain a session cookie. We can log in using the built-in login controller. To log in, simply send a post request to the .../login path with the correct username and password:

```
curl -v "POST" "http://localhost:8081/login" -H
'Content-Type: application/json; charset=utf-8' -d
'{"username":"alice","password":"alice@1"}'
```

If the request succeeds, a cookie will be sent in the response:

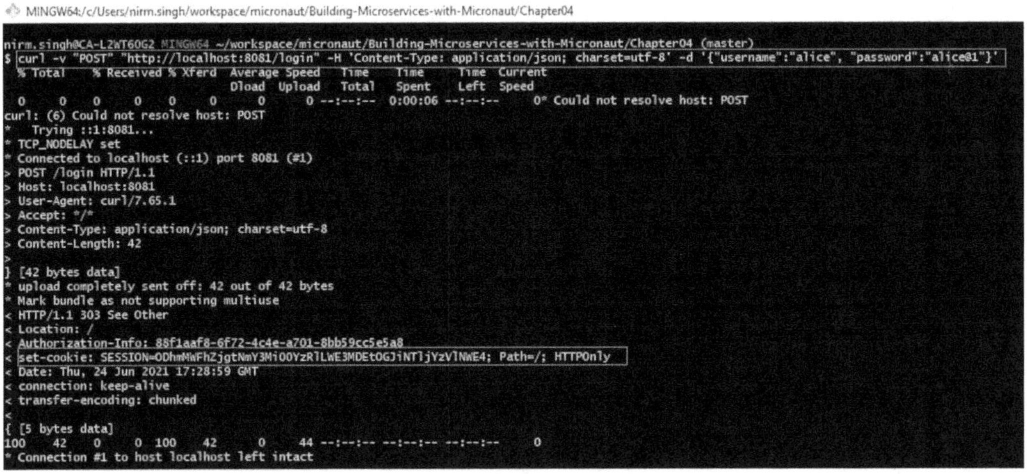

Figure 4.7 – Obtaining a cookie for the secured access

As observed in the preceding screenshot, we will send a post request to .../login using the correct username and password, to which the service will return a cookie in response.

We can pass this cookie to any requests to `OwnerResource`. In the following screenshot, we passed the obtained cookie to make an HTTP GET call to the `/api/owners` endpoint:

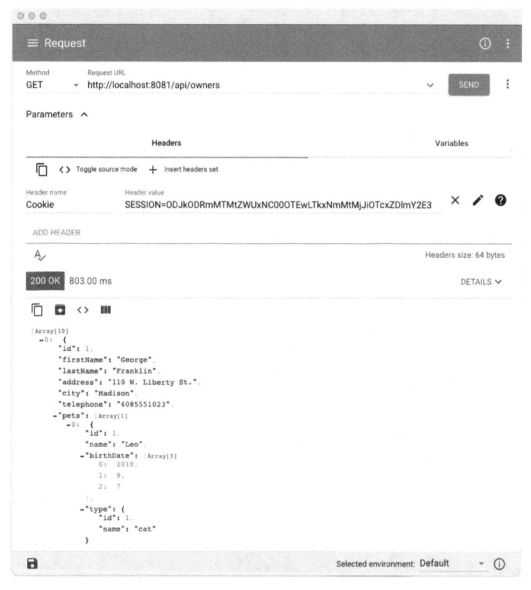

Figure 4.8 – Using an obtained cookie for the secured access

Since we passed the obtained cookie in the request headers, the service will extract this cookie, validate, and successfully return the HTTP 200 response.

Until now, we covered how to address anonymous and authenticated access scenarios using session authentication. In the next section, we will dive into using **JSON Web Tokens (JWTs)** for securing access to a microservice in the Micronaut framework.

# Using JWT authentication to secure the service endpoints

In token-based authentication, the user state is stored at the client side. When a client logs in to the server, the server encrypts the user data into a token with a secret and sends it back to the client. Any subsequent user requests must have this token set in the request header. The server retrieves the token, validates the authenticity, and resumes the user session:

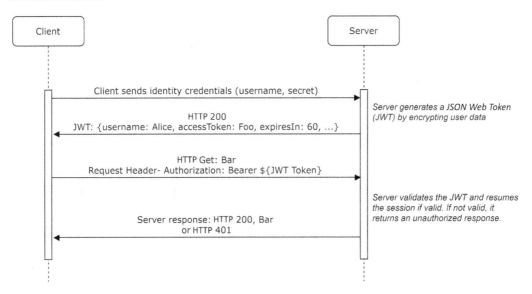

Figure 4.9 – Token-based authentication

As shown in the preceding diagram, in a token-based authentication strategy, the client does the heavy lifting of keeping track of the session in the JSON web token. A client must provide a valid token to resume the session.

To learn how to secure a microservice using token-based authentication, we will work on a hands-on `pet-clinic` microservice. To begin, we will set up a third-party identity provider using Keycloak. In the next section, we will set up Keycloak locally.

# Setting up Keycloak as the identity provider

We will run the Keycloak server in the local Docker container. If you don't have Docker installed, you may refer to the *Technical requirements* section to see how to install Docker on your development workspace. To boot up a local Keycloak server in Docker, open a Bash terminal and run the following command:

```
$ docker run -d --name keycloak -p 8888:8080 -e KEYCLOAK_
USER=micronaut -e KEYCLOAK_PASSWORD=micronaut123 jboss/keycloak
```

After this, Docker will instantiate a Keycloak server in a container and mount container port `8080` to host operating system port `8888`. Furthermore, it will create a `micronaut` admin user with the password as `micronaut123`. After successful installation, you can access Keycloak at `http://localhost:8888/`. In the next section, we will begin by setting up a client for the microservice.

## Creating a client on the Keycloak server

To use Keycloak as an identity provider, we will start with setting up a client. Follow these instructions to set up a Keycloak identity provider client:

1. Access the **Keycloak admin** module at `http://localhost:8888/auth/admin/`.

2. Provide a valid admin username and password (in our case, it's `micronaut` and `micronaut123`).

3. Select **Clients** from the left navigation menu.

4. Provide a client ID (you can skip the rest of the inputs).

5. After creating a client with the provided Client ID, Keycloak will open the settings tab for the client. You must select the highlighted values shown in the following screenshot:

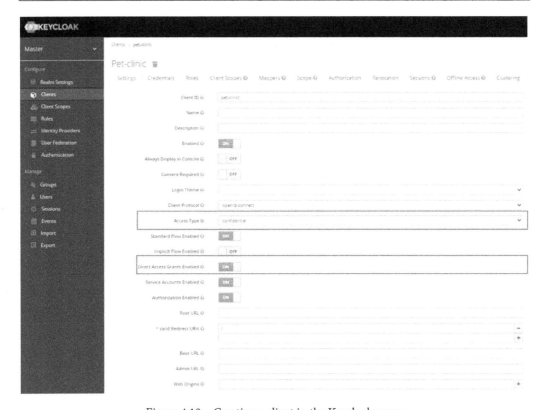

Figure 4.10 – Creating a client in the Keycloak server

The Keycloak server will create the `pet-clinic` client in the default master realm. Next, we will set up some users for this client space.

## Setting up the users in the client space

The user setup will enable us to use these identities as test users (and, of course, later, the actual users can be configured). We will begin by creating the roles. For the `pet-clinic` microservice, we will define two roles: `pet-clinic-admin` and `pet-clinic-user`. To create a role, follow the instructions mentioned next:

1.  Select **Roles** on the main menu.
2.  Hit the **Add Role** button.
3.  Provide a role name and hit the **Save** button.

We will add three users – Alice (admin), Bob (user), and Charlie (user). To add a user, follow the instructions mentioned next:

1.  Select **Users** on the main menu and hit **Add User**.
2.  Provide a username and keep the default settings. Hit the **Save** button.
3.  Once the user is created, go to the **Credentials** tab specify the password and change the **Temporary** flag to **off**. Hit the **Reset Password** button.
4.  To configure user-role, go to the **Role Mappings** tab and select the desired user role. Changes will be saved automatically.

Repeat the preceding instructions to set up Alice as pet-clinic-admin, Bob as pet-clinic-user, and Charlie as pet-clinic-user.

In order to surface this role data from Keycloak, we need to make the following changes:

1.  Select the **Client Scopes** option from the main menu.
2.  Select **Roles** in the listed options.
3.  Go to the **Mappers** tab for the roles and select **Realm Roles**.
4.  Provide the input as highlighted in the following screenshot:

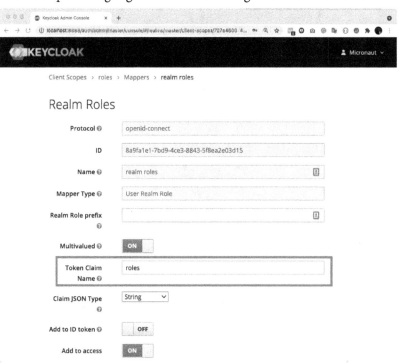

Figure 4.11 – Configuring the Realm Roles mapper

Once the users are created and roles are assigned, we can proceed with the `pet-clinic` microservice changes. In the next section, we will dive into making `pet-clinic` secure using token-based authentication with a Keycloak identity provider.

## Securing the pet-clinic microservice using token-based authentication

To secure the `pet-clinic` microservice, we will first need to enable security by adding the following dependencies in the `pom.xml` project:

```xml
<!-- Micronaut security -->
    <dependency>
        <groupId>io.micronaut.security</groupId>
        <artifactId>micronaut-security</artifactId>
        <version>${micronaut.version}</version>
    </dependency>
    <dependency>
        <groupId>io.micronaut.security</groupId>
        <artifactId>micronaut-security-jwt</artifactId>
        <version>${micronaut.version}</version>
    </dependency>
    <dependency>
        <groupId>io.micronaut.security</groupId>
        <artifactId>micronaut-security-oauth2</artifactId>
        <version>${micronaut.version}</version>
    </dependency>
...
```

By importing the `micronaut-security` and `micronaut-security-jwt` dependencies, we can leverage the token authentication toolkit in the `pet-clinic` microservice. We will use OAuth 2 for integrating with the Keycloak server. Once these dependencies are imported, we will then need to configure `application.properties` as follows:

```
security:
    authentication: idtoken
    endpoints:
      login:
        enabled: true
```

```
    redirect:
      login-success: /secure/anonymous
    token:
      jwt:
        enabled: true
        signatures.jwks.keycloak:
          url: http://localhost:8888/auth/realms/master/
protocol/openid-connect/certs
    oauth2.clients.keycloak:
      grant-type: password
      client-id: pet-clinic
      client-secret: XXXXXXXXX
      authorization:
        url: http://localhost:8888/auth/realms/master/protocol/
openid-connect/auth
      token:
        url: http://localhost:8888/auth/realms/master/protocol/
openid-connect/token
        auth-method: client_secret_post
```

In the application properties, `client-id` and `client-secret` must be copied from `KeyCloak`. Client secret can be copied by going to **Clients** | **pet-clinic** client | **Credentials** tab. Furthermore, the URLs for authorization and tokens are standard but you can access all the configurations at `http://localhost:8888/auth/realms/master/.well-known/openid-configuration`.

In the next section, we will focus on how to grant secured access to the controller endpoints using the configured token-based authentication strategy and Keycloak identity server.

## Granting secured access using the KeyCloak identity provider

To grant secured access, we can use the `@Secured` annotation as well as `intercept-url-map`. In the hands-on example, we will grant secured access to `VetResource` using the `@Secured` annotation, as shown in the following code snippet:

```
@Controller("/api")
@Secured(SecurityRule.IS_AUTHENTICATED)
public class VetResource {
...
}
```

All the endpoints within `VetResource` are granted only secured access. If we try to hit any .../`vets` endpoints, the microservice will return a forbidden response:

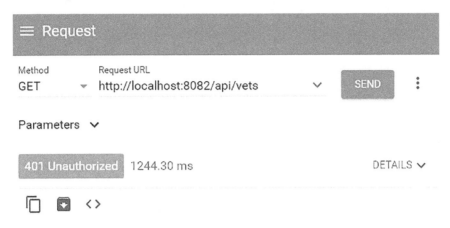

Figure 4.12 – Unauthenticated access to vets

As shown in the preceding figure, if we try to access any vet endpoint without specifying a valid token, the microservice will throw an `HTTP 401 Unauthorized` response.

For successful access to the vet endpoints, we will need to obtain a valid JWT. We can log in using the built-in login controller. To log in, simply send a post request to the .../`login` path with the correct username and password. If the request succeeds, a cookie with a JWT will be sent in the response, as shown in the next screenshot:

Figure 4.13 – Obtaining a cookie for the secured access

As highlighted in the screenshot, we will send a post request to the .../login endpoint using the correct username and password. The service will grant the secured access using the KeyCloak identity provider and return a cookie with the JWT.

Copy the JWT portion from the preceding response. We can pass this token to any requests to VetResource. In the following screenshot, we are invoking the .../vets endpoint using the just-obtained JWT:

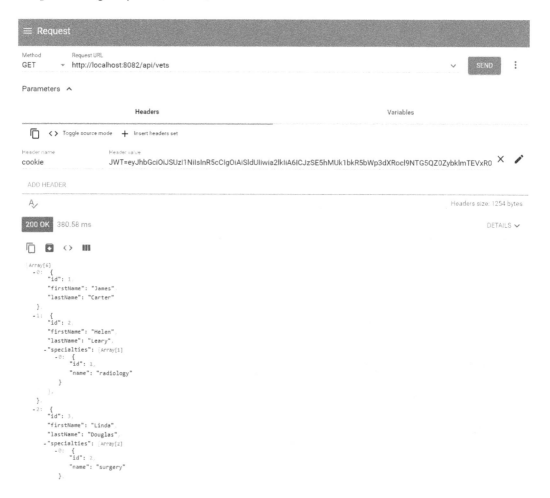

Figure 4.14 – Using the obtained token for the secured access

Since we passed a valid token in the request headers, the service will validate this token and successfully return the HTTP 200 response.

So far, we have explored how to secure a microservice using a JWT with an external identity provider. In the next section, we will focus on how to implement microservice security using OAuth with a cloud identity provider.

# Using OAuth to secure service endpoints

OAuth is yet another token-based authentication strategy. Its wide acceptability, good coverage of the depth and breadth of web security concerns, and the flexibility of managing user sessions at both the client and server side make it an enterprise-grade authentication mechanism. OAuth dictates to use the token to establish the identity instead of passing usernames and passwords. A token can be obtained from an external identity provider and this token can then be passed to any subsequent requests to resume a session:

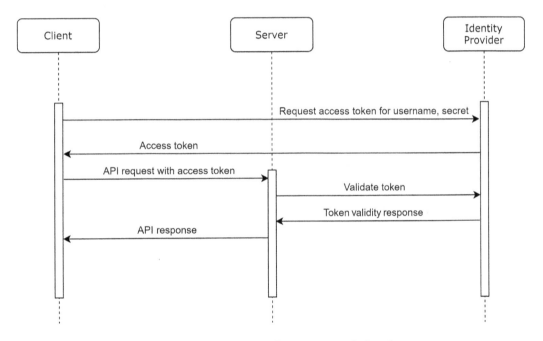

Figure 4.15 – Separating the concerns with OAuth

As shown in the preceding figure, in the OAuth token-based authentication strategy, the client obtains a token from the identity provider and uses this token in any API requests to the server. The server validates this token with the identity provider to return a proper response.

To learn how to secure a microservice using an OAuth and cloud-based identity provider, we will do a hands-on exercise with the `pet-clinic-review` microservice. To begin, we will set up a cloud identity provider using Okta.

# Setting up Okta as the identity provider

Okta is a leading SaaS identity management portal. We will use Okta as the identity provider. In order to begin, you must be registered with Okta. Sign up at `developer.okta.com`. Once you're signed up, Okta will ask the user to confirm the email. In the email, you will also receive an Okta domain, as highlighted in the following screenshot:

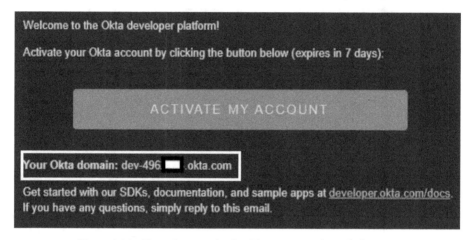

Figure 4.16 – Domain name in the Okta signup acknowledgment

As shown in the figure, an Okta domain will be created for your developer account. You must save this as this will be used later to configure Okta as the identity provider.

In the next section, we will see how to create an application on Okta.

## Creating an app on Okta

In order to use Okta with your microservice, you are required to create an app on Okta. Follow the instructions given as follows for creating an app on Okta:

1. Log on to `https://developer.okta.com/`.
2. On the landing page, select **Creating a Web Application**.
3. Choose **Native** as your platform and hit the **Next** button.

4.  Provide the app settings mentioned in the following screenshot and once all the inputs are provided, hit the **Done** button:

APPLICATION SETTINGS

| | |
|---|---|
| Name | pet-clinic-reviews |
| Login redirect URIs | com.okta.dev-4962048:/callback  × |
| | + Add URI |
| | Okta sends an OAuth authorization response to these URIs. Add your application's callback endpoint. Docs |
| Logout redirect URIs | com.okta.dev-4962048:/  × |
| | + Add URI |
| | When a user signs out, your application can specify a URI where the browser is redirected. Okta only allows redirects for URIs that are listed here. Docs |
| Group assignments Optional | ◉ Everyone × |
| | Users can only sign in to apps that they are assigned to. Group assignments are easier to manage than individual users. |
| Grant type allowed | Client acting on behalf of a user |
| | ☑ Authorization Code |
| | ☑ Refresh Token |
| | ☑ Resource Owner Password |
| | ☑ Implicit (Hybrid) |
| | Okta can authorize your app's requests with these OAuth 2.0 grant types. Limit the allowed grant types to minimize security risks. Docs |

Figure 4.17 – Creating the microservice app on Okta

5.  We will keep most of the inputs as their defaults. Under **Grant type allowed**, check all the boxes.

6.  Once the app is created successfully, edit **Client Credentials** and select the **Use client authentication** option for **Client authentication**.

After creating the application as previously instructed, jot down the client ID and client secret. This will be used later. Next, we will set up some users for this app.

## Setting up the users in the client space

The user setup will enable us to use these identities as test users. We will add three users – Alice (admin), Bob (user), and Charlie (user). To add a user, follow the given instructions:

1.  On the main navigation bar, hover over **Users** and select **People**.

2.  Hit the **Add Person** button.

3.  Provide inputs as shown in the following screenshot:

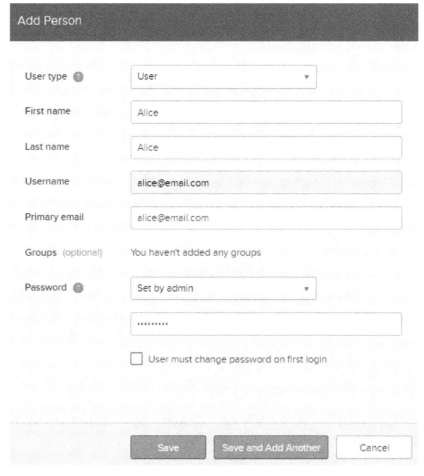

Figure 4.18 – Adding a person (user) on Okta

4.  In the password input, choose **Set by admin** and you must keep **User must change password on first login** unchecked. This will allow us to quickly use the identity without resetting the password.

Repeat the preceding instructions for setting up `Bob` and `Charlie` as application users. Once the users are created, we can proceed with the `pet-clinic-reviews` microservice changes.

In the next section, we will dive into making `pet-clinic-reviews` secure but let's first start with enabling SSL to make encrypted communication over HTTPS.

# Enabling SSL in the Micronaut framework

Any security safeguarding is left incomplete if a microservice is not exposed to HTTPS. In earlier sections, we purposely focused on authentication and authorization only while skipping SSL. As we will be using a third-party identity provided over the cloud, it's recommended and required to enable SSL in the `pet-clinic-reviews` microservice.

In order to enable SSL, we will need an SSL certificate for localhost. We will be creating a self-signed certificate using OpenSSL. Follow these instructions for creating a self-signed certificate using OpenSSL:

1.  Open a Git Bash terminal.

2.  Change directory to the root directory of the `pet-clinic-reviews` project.

3.  Run `winpty openssl req -x509 -newkey rsa:2048 -keyout key.pem -out cert.pem -days 365` in Git Bash. Provide the correct information to create a self-signed certificate. This will create a `key.pem` file and `cert.pem` file in the opened directory.

4.  To combine the key and certificate files, run `winpty openssl pkcs12 -inkey key.pem -in cert.pem -export -out cert.p12` in Git Bash.

5.  To verify that you've created the P12 file, run `winpty openssl pkcs12 -in cert.p12 -noout -info` in Git Bash. You must provide the same password that was used to create the P12 file.

Following the previous instructions, we can successfully create a platform-agnostic certificate. The P12 format has gained popularity as it can be used across platforms and operating systems. Next, we will add this certificate to the host operating system trust store so it can be trusted by all the running applications on the system. Follow the instructions mentioned next to add the certificate to the trust store:

1.  Determine `$JAVA_HOME`. It can be found in the system variables.

2.  Copy the just-created `cert.pem` file to `$JAVA_HOME/jre/lib/security/cacerts`.

3.  Open the Git Bash terminal with admin rights and change the directory to `$JAVA_HOME/jre/lib/security`.

4.  Run `winpty keytool -importcert -file cert.pem -alias localhost -keystore $JAVA_HOME/jre/lib/security/cacerts -storepass changeit` in the Git Bash terminal.

By following the previous instructions, we will add the self-signed certificate to the trust store. This will enable the system to trust this certificate when it's used over SSL.

Our custom developer Okta domain may also not be trusted by the system. We will follow similar instructions to add the Okta certificate to the `cacerts` trust store:

1.  Open a new tab in the Chrome browser. Open the developer tools.

2.  Hit `https://${yourOktaDomain}/oauth2/default/.well-known/oauth-authorization-server?client_id=${yourClientId}`.

3.  In the developer tools, go to the **Security** tab and click on **View certificate**.

4.  This will open the certificate in a prompt. Go to the **Details** tab on this prompt.

5.  Click on **Copy file** and follow the instructions to export the certificate to a local directory.

6.  Copy the just-exported certificate to `$JAVA_HOME/jre/lib/security/cacerts`.

7.  Open the Git Bash terminal with admin rights and change the directory to `$JAVA_HOME/jre/lib/security`.

8.  Run `winpty keytool -importcert -file okta.cert -alias localhost -keystore $JAVA_HOME/jre/lib/security/cacerts -storepass changeit` in the Git Bash terminal. In the file option, you must provide the exported certificate name.

Adding the developer domain Okta certificate to the system trust store will enable us to communicate with the Okta identity provider. In the next section, we will dive into enabling SSL in the `pet-clinic-reviews` microservice using the self-signed certificate.

## Configuring the application properties for SSL

Once you have a legible certificate, the Micronaut framework provides a quick way to turn on SSL by configuring some application properties. Make the following changes to the application properties to enable SSL:

```
micronaut:
  ssl:
    enabled: true
    key-store:
      type: PKCS12
      path: file:cert.p12
      password: Pass@w0rd
```

To enable SSL, we have used a self-signed certificate that we created in the previous section. The **password** field must match the password used to create the certificate. By making these changes to the application properties, Micronaut will enable SSL for the application and use port 8443 for secured communication.

In the next section, we will focus on how to configure the pet-clinic-reviews microservice with OAuth security using the Okta identity provider.

# Securing the pet-clinic-reviews microservice using OAuth

To secure the pet-clinic microservice, we will first need to enable security by adding the following dependencies in the pom project:

```
<!-- Micronaut security -->
    <dependency>
        <groupId>io.micronaut.security</groupId>
        <artifactId>micronaut-security</artifactId>
    </dependency>
    <dependency>
        <groupId>io.micronaut.security</groupId>
        <artifactId>micronaut-security-jwt</artifactId>
    </dependency>
    <dependency>
        <groupId>io.micronaut.security</groupId>
        <artifactId>micronaut-security-oauth2</artifactId>
```

```
        </dependency>
    ...
```

By importing the `micronaut-security` and `micronaut-security-jwt` dependencies, we can leverage the token authentication and OAuth toolkit in the `pet-clinic-reviews` microservice. Once these dependencies are imported, we will then need to configure `application.properties` as follows:

```
security:
    authentication: idtoken
    oauth2:
      clients:
        okta:
          client-secret: HbheS-
            q4P6oewQgT7uK58bgMbtHbCwcarzWuHB32
          client-id: 0oa37vkb7Sq23P1kh5d6
          openid:
            issuer: https://dev-
              4962048.okta.com/oauth2/default
    endpoints:
      logout:
        get-allowed: true
```

In the application properties, `client-id` and `client-secret` must be copied from Okta. For the issuer, you must provide your Okta domain in the first part. You might just need to change your developer domain but you can get more information on authorization and token URLs by accessing the OAuth configurations at `https://${yourOktaDomain}/oauth2/default/.well-known/oauth-authorization-server?client_id=${yourClientId}`.

In the next section, we will focus on how to grant secured access to the controller endpoints using the OAuth and Okta identity servers.

## Granting secured access using the Okta identity provider

For granting secured access, we can use the `@Secured` annotation as well as `intercept-url-map`. In our hands-on example, we will grant secured access to `VetReviewResource` using the `@Secured` annotation:

```
@Controller("/api")
@Secured(SecurityRule.IS_AUTHENTICATED)
```

```
public class VetReviewResource {

    ...

}
```

All the endpoints within `VetReviewResource` are granted only secured access. If we try to hit any `.../vet-reviews` endpoints, the microservice will return a forbidden response. In the following figure, we tried to access the `.../vet-reviews` endpoint unsecured and the service responded with `HTTP 401`:

```
nirm.singh@CA-L2WT60G2 MINGW64 ~/workspace/micronaut/chapter-04/micronaut-petclinic/pet-clinic-reviews
$ curl -k -v GET "https://localhost:8443/api/vet-reviews"
  % Total    % Received % Xferd  Average Speed   Time    Time     Time  Current
                                 Dload  Upload   Total   Spent    Left  Speed
  0     0    0     0    0     0      0      0 --:--:--  0:00:01 --:--:--     0* Could not resolve host: GET
curl: (6) Could not resolve host: GET
*   Trying ::1:8443...
* TCP_NODELAY set
* Connected to localhost (::1) port 8443 (#1)
* ALPN, offering h2
* ALPN, offering http/1.1
* successfully set certificate verify locations:
*   CAfile: C:/Program Files/Git/mingw64/ssl/certs/ca-bundle.crt
  CApath: none
} [5 bytes data]
* TLSv1.3 (OUT), TLS handshake, Client hello (1):
} [512 bytes data]
* TLSv1.3 (IN), TLS handshake, Server hello (2):
{ [85 bytes data]
* TLSv1.2 (IN), TLS handshake, Certificate (11):
{ [1053 bytes data]
* TLSv1.2 (IN), TLS handshake, Server key exchange (12):
{ [300 bytes data]
* TLSv1.2 (IN), TLS handshake, Server finished (14):
{ [4 bytes data]
* TLSv1.2 (OUT), TLS handshake, Client key exchange (16):
} [37 bytes data]
* TLSv1.2 (OUT), TLS change cipher, Change cipher spec (1):
} [1 bytes data]
* TLSv1.2 (OUT), TLS handshake, Finished (20):
} [16 bytes data]
  0     0    0     0    0     0      0      0 --:--:-- --:--:-- --:--:--     0* TLSv1.2 (IN), TLS handshake, Finished (20):
{ [16 bytes data]
* SSL connection using TLSv1.2 / ECDHE-RSA-AES128-GCM-SHA256
* ALPN, server did not agree to a protocol
* Server certificate:
*  subject: C=CA; ST=Ontario; L=Toronto; O=Packtpub; OU=Packtpub; CN=Nirmal Singh; emailAddress=singhnirmal90@gmail.com
*  start date: Dec 30 20:43:16 2020 GMT
*  expire date: Dec 30 20:43:16 2021 GMT
*  issuer: C=CA; ST=Ontario; L=Toronto; O=Packtpub; OU=Packtpub; CN=Nirmal Singh; emailAddress=singhnirmal90@gmail.com
*  SSL certificate verify result: self signed certificate (18), continuing anyway.
} [5 bytes data]
> GET /api/vet-reviews HTTP/1.1
> Host: localhost:8443
> User-Agent: curl/7.65.1
> Accept: */*
>
{ [5 bytes data]
* Mark bundle as not supporting multiuse
< HTTP/1.1 401 Unauthorized
< connection: keep-alive
< transfer-encoding: chunked
<
{ [5 bytes data]
  0     0    0     0    0     0      0      0 --:--:-- --:--:-- --:--:--     0
* Connection #1 to host localhost left intact

nirm.singh@CA-L2WT60G2 MINGW64 ~/workspace/micronaut/chapter-04/micronaut-petclinic/pet-clinic-reviews
$ |
```

Figure 4.19 – Unauthenticated access to vets

As highlighted in the previous screenshot, if we try to access any `vet-reviews` endpoint without specifying a valid token, the microservice will throw an `HTTP 401 Unauthorized` response.

For successful access to the vet endpoints, we will need to obtain a valid JWT. We can obtain a valid token by accessing the Okta token API. The following is the `curl` command to call the Okta token API:

```
curl -k -u client_id:client_secret \
--location --request POST 'https://dev-4962048.okta.com//
oauth2/default/v1/token' \
--header 'Accept: application/json' \
--header 'Content-Type: application/x-www-form-urlencoded' \
--data-urlencode 'grant_type=password' \
--data-urlencode 'username=Alice' \
--data-urlencode 'password=Pass@w0rd' \
--data-urlencode 'scope=openid'
```

In the previous `curl` command, you must provide the correct values in `client_id` and `client_secret`, the `POST` URL and the user credentials. If everything is validated successfully, the token API will respond back with a bearer and ID token. Copy the returned access token. We can pass this token to any requests to `VetReviewResource` for secured communication:

Figure 4.20 – Using the obtained token for secured access

Since we passed a valid token in the request headers, the service will validate this token and successfully return the `HTTP 200` response.

In this section, we learned about and experimented with OAuth security. To make `pet-clinic-reviews` service endpoints secure, we used OAuth with Okta as a third-party identity provider.

# Summary

In this chapter, we explored various ways to secure microservices in the Micronaut framework. We began our journey by diving into a session authentication strategy, and then we explored token-based authentication using an external Keycloak identity server. Lastly, we worked on securing a microservice using OAuth with a cloud-based identity provider. Furthermore, we also worked on enabling SSL to make service communication secure over HTTPS.

This chapter provided you with a handy, focused skillset in safeguarding a microservice in the Micronaut framework using various authentication strategies, along with how to work with local or external (cloud) identity providers.

In the next chapter, we will explore how we can integrate different microservices using event-driven architecture.

# Questions

1. What are the various authentication strategies in the Micronaut framework?

2. What is a security filter in Micronaut?

3. How do you set up session-based authentication in the Micronaut framework?

4. What is the `@Secured` annotation in the Micronaut framework?

5. What is `intercept-url-maps` in the Micronaut framework?

6. How do you set up token-based authentication in the Micronaut framework?

7. How do you set up JWT authentication in the Micronaut framework?

8. How do you integrate with Keycloak in the Micronaut framework?

9. How do you set up OAuth authentication in the Micronaut framework?

10. How do you integrate with Okta in the Micronaut framework?

11. How do you enable SSL in the Micronaut framework?

# 5

# Integrating Microservices Using Event-Driven Architecture

The essence of microservice architecture is breaking a monolith down into decoupled or loosely coupled microservices. As a result of such a decomposition into microservices, we separate the user and/or business concerns owned by each microservice. However, for an application as a whole, all the microservices need to work together by interacting with each other in executing and serving the user requests. Event-driven architecture has gained popularity in addressing these inter-microservices interactions.

In this chapter, we will explore how we can implement an event-driven architecture in the Micronaut framework. We will dive into the following topics:

- Understanding event-driven architecture
- Event streaming with the Apache Kafka ecosystem
- Integrating microservices using event streaming

By the end of this chapter, readers will have a nifty knowledge of event-driven architecture and how to implement an event-streaming broker to integrate app microservices in the Micronaut framework.

# Technical requirements

All the commands and technical instructions in this chapter are run on Windows 10 and macOS. Code examples covered in this chapter are available in the book's GitHub repository at `https://github.com/PacktPublishing/Building-Microservices-with-Micronaut/tree/master/Chapter05`.

The following tools need to be installed and set up in the development environment:

- **Java SDK**: Java SDK version 13 or above (we used Java 14).

- **Maven**: This is optional and only required if you would like to use Maven as the build system. However, we recommend having Maven set up on any development machine. Instructions regarding the downloading and installation of Maven can be found at `https://maven.apache.org/download.cgi`.

- **Development IDE**: Based on your preferences, any Java-based IDE can be used, but for purposes of this chapter, IntelliJ was used.

- **Git**: Instructions regarding the downloading and installation of Git can be found at `https://git-scm.com/downloads`.

- **PostgreSQL**: Instructions regarding downloading and installation can be found at `https://www.postgresql.org/download/`.

- **MongoDB**: MongoDB Atlas provides a free online database-as-a-service up to 512 MB storage. However, if a local database is preferred, then instructions regarding downloading and installation can be found at `https://docs.mongodb.com/manual/administration/install-community/`. We used a local installation while writing this chapter.

- **REST client**: Any HTTP REST client can be used. We used the **Advanced REST Client (ARC)** Chrome plugin.

- **Docker**: Instructions regarding the downloading and installation of Docker can be found at `https://docs.docker.com/get-docker/`.

# Understanding event-driven architecture

Event-driven architecture is pivotal in connecting different microservices. Before we dive into how to implement an event-driven interaction system, let's understand its fundamentals.

The following are the key components at the core of any event-driven architecture implementation:

- **Event**: An event is simply a change in the state of the system that needs to be traced. In a microservice architecture, a microservice may make or detect a change in the data's state that might be worth noticing by other services. This state change is communicated as an event.

- **Event producer**: An event producer is any microservice or component that is making or detecting a state change and generating an event for other components/ services in the system.

- **Event consumer**: An event consumer is any microservice or component that is consuming an event. Interestingly, this event consumption might trigger this component to produce another event.

- **Event broker**: The event broker acts as a go-between between all the producer and consumer parties. It maintains a metadata quorum to keep track of events.

These key components come together to realize an event-driven architecture. Broadly speaking, there are two implementation strategies – **pub/sub** (also called event messaging) and **event streaming**. To learn more about these strategies, let's dive into the following sections.

## Event messaging or a pub/sub model in an event-driven architecture

A **pub/sub** model is a *push-based* model. In a push-based model, event publishing is owned by the event producer, and events are pushed from the producer and sent to consumers. The following are the key components in a pub/sub implementation:

- **Event producer**: Any component that is making or detecting a state change will generate an event and publish it to the event broker.

- **Event broker**: The event broker will receive the generated event and push the event to all the required event queues. These event queues are subscribed to by event consumers. Therefore, events are pushed down to the event consumers by the broker.

- **Event consumer**: The event consumer will receive the event and do what is required. It may also generate a new event(s).

  This is depicted in the following diagram:

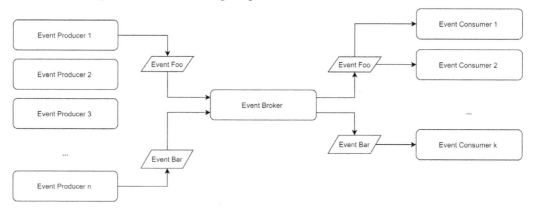

Figure 5.1 – Pub/sub model

As shown in the preceding diagram, when **Event Producer 1** generates the **Event Foo**, it is pushed to the **Event Broker**. The **Event Broker** further pushes this event to **Event Consumer 1** and **Event Consumer 2**.

So, overall, for **Event Bar** (which is generated by **Event Producer n**), **Event Broker** pushes it to **Event Consumer k**.

In a pub/sub model, once the event is produced and communicated to the consumer via the event broker, the event consumer must do the necessary immediately, as once an event is consumed, it perishes. The event consumer can never go back to historic events. This model is also sometimes referred to as the **event messaging model**.

## Event streaming in event-driven architecture

An **event streaming** model is a *pull-based* model. In a pull-based model, the onus lies on the event consumer to fetch the event. In an event streaming implementation, the key components will act as follows:

- **Event producer**: Any component that is making or detecting a state change will generate the event and send it to the event broker.

- **Event broker**: The event broker will receive the generated event and broadcast the event by putting the event in an event stream.

- **Event consumer**: The event consumer continuously monitors one or more event streams on the event broker. When a new event is pushed to the event stream, the consumer fetches the event and does what is required.

Refer to the following diagram:

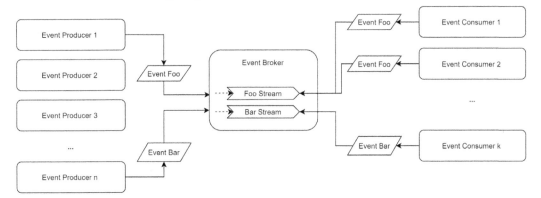

Figure 5.2 – Event streaming model

As shown in the preceding diagram, when **Event Producer 1** generates the **Event Foo**, it pushes the event to **Event Broker** and **Event Broker** puts it in the **Foo Stream**. **Event Consumer 1** and **Event Consumer 2** fetch the event from the **Foo Stream**. **Event Bar** is fetched from **Bar Stream** by **Event Consumer k**.

In an event-streaming model, as event consumers fetch the data from an event stream, they can fetch any offset of the event stream. This even enables event consumers to access historic events. It especially comes in handy if you have a new consumer added to the system that might not be in touch with the recent state of the system and may start processing historic events first. For these reasons, event streaming is usually preferred over event messaging.

In the next section, we will get started with hands-on event streaming using a popular event-streaming stack.

# Event streaming with the Apache Kafka ecosystem

Apache Kafka is an industry-leading event streaming system. In the Apache Kafka ecosystem, the following are some of the key components:

- **Event topic**: An event topic consists of a stream of immutable, ordered messages belonging to a particular category. Each event topic may have one or more partitions. A partition is indexed storage that supports multi-concurrency in Apache. Apache Kafka keeps at least one partition per topic and may add more partitions as specified (at the time of topic creation) or required. When a new message is published to the topic, Apache Kafka decides which topic partition will be used to append the message. Each topic appends the most recent message at the end. This is shown in the following diagram:

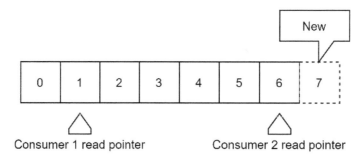

Figure 5.3 – Apache Kafka topic anatomy

As shown in the preceding diagram, when a new message is published to the steam, it is appended at the end. Event consumers can freely choose which topic offset to read. While **Consumer 1** reads at the first offset, **Consumer 2** reads from the sixth offset.

- **Event broker**: An event broker is a façade that provides an interface to write or read from event topic(s). Apache Kafka usually has the leader and follower brokers. The leader broker (for a topic) will serve all the write requests. If a leader broker fails, then the follower broker chimes in as leader.

- **Kafka cluster**: If Apache Kafka has a quorum consisting of more than one event broker, then it's called a cluster. In a cluster, each broker will usually lead a distinct topic and may substitute as a follower for other topics.

- **Event producer**: Producers publish an event message to the topic stream. A producer interacts with the Apache Kafka ecosystem to know which broker should be used for writing an event message to a topic stream. If a broker fails or a new broker is added, Apache Kafka notifies the required producers of this.

- **Event consumer**: A consumer reads the event messages from the topic stream. An event consumer will interact with the Apache Kafka ecosystem to know which broker should be used to read from a topic stream. Furthermore, Apache Kafka keeps track of the topic offset for each event consumer to resume event consumption properly.

- **Zookeeper**: Zookeeper maintains the metadata quorum for the Apache Kafka ecosystem. It essentially maintains information about all brokers for event producers and consumers. It also keeps track of the topic offset for each event consumer.

In the following diagram, we can see the various components in the Apache Kafka ecosystem and how they interact with each other in event streaming:

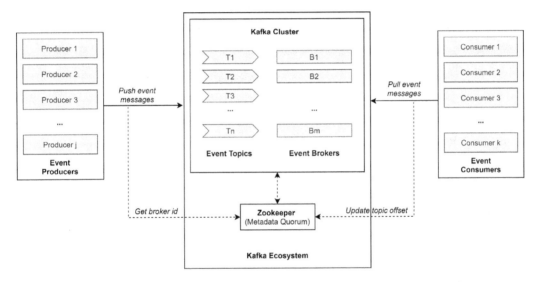

Figure 5.4 – Apache Kafka ecosystem

In the preceding diagram, the Apache Kafka ecosystem is shown at a glance. For each event topic, there will be at least one leader event broker and one or more follower event brokers. Information about leaders and followers is maintained in Zookeeper. When an event producer pushes a message, a broker will write the message to the required topic stream. Similarly, when an event consumer pulls a message, an event broker will fetch the message from the required topic stream. Zookeeper maintains the offset information for each topic for all the event consumers.

In the next section, we will dive into how to use the Apache Kafka ecosystem for event streaming in the pet clinic application.

# Integrating microservices using event streaming

To learn and perform hands-on exercises, we will implement a simple scenario of event streaming in the pet clinic application. Consider the following diagram:

Figure 5.5 – Event streaming in the pet clinic application

In the aforementioned diagram, whenever there is a new vet review, the **pet-clinic-reviews** microservice will send the review to **Apache Kafka Streaming**. Apache Kafka appends the review to the **vet-reviews** topic stream. And, as the **pet-clinic** microservice is continuously monitoring the **vet-reviews** topic stream, it will fetch any new reviews appended to the topic and update the average rating accordingly. This is a simpleton diagram but will help to focus on the key learning objectives.

In the next section, we will begin by setting up the Apache Kafka ecosystem locally inside Docker to learn more about Apache Kafka streaming.

## Setting up the Apache Kafka ecosystem locally

To set up the Apache Kafka ecosystem locally, we will use Docker. The `docker-compose` file for all the required components and configurations can be found in the chapter's GitHub workspace under `resources`: `https://github.com/PacktPublishing/Building-Microservices-with-Micronaut/blob/master/Chapter05/micronaut-petclinic/pet-clinic-reviews/src/main/resources/kafka-zookeeper-kafdrop-docker/docker-compose.yml`.

Perform the following steps to install and set up Apache Kafka:

1. Download the `docker-compose` file from the aforementioned URL.

2. Open the GitBash terminal.

3. Change the directory to where you have placed the `docker-compose` file.

4. Run the `docker-compose up` command in the GitBash terminal.

As a result of following these instructions, Docker will install Zookeeper, Apache Kafka, and Kafdrop. Kafdrop is an intuitive admin GUI for managing Apache Kafka. In the following section, we will verify their installation.

## Testing the Apache Kafka ecosystem setup

To test whether the Apache Kafka ecosystem is installed successfully, perform the following steps:

1.  Open the GitBash terminal and run the following command:

    ```
    winpty docker exec -it kafka bash
    ```

2.  Change the directory to **opt/bitnami/kafka/bin/**.

3.  Add a topic stream by running the following in the GitBash terminal:

    ```
    command ./kafka-topics.sh --bootstrap-server kafka:9092
    --create --partitions 3 --replication-factor 1 --topic
    foo-stream
    ```

4.  To add a message to the topic, run the following in the GitBash terminal:

    ```
    command ./kafka-console-producer.sh --broker-list
    kafka:9092 --topic foo-stream
    ```

5.  A terminal prompt will appear, type `hello-world!`, and then hit *Enter*.

6.  Press *Ctrl + D*, which should successfully add the event to the topic.

By following these instructions, we added a `foo-stream` topic and added a message to this topic. To see this topic stream, we can open Kafdrop by opening `http://localhost:9100/` in a browser window. Refer to the following screenshot:

Figure 5.6 – Kafdrop showing foo-stream topic messages

Kafdrop provides an intuitive GUI for viewing and managing all the Apache Kafka streams. In the previous screenshot, we can see the messages inside the just created **foo-stream**.

Hitherto, we set up the Apache Kafka ecosystem locally in a Dockerized environment, and in the next section, we will use this setup for hands-on event streaming in the `pet-clinic-reviews` and `pet-clinic` microservices. We will begin by making the required changes in the `pet-clinic-reviews` microservice.

## Implementing an event-producer client in the pet-clinic-reviews microservice

We will begin making the required changes to the `pet-clinic-reviews` microservices so that it can stream out the vet reviews to Apache Kafka. For this hands-on exercise, we will keep things simple. Therefore, we will skip the security setup and resume the code base from *Chapter 3, Working on the RESTful Web Services*.

Perform the following steps to see how this goes:

1. To begin, we will need to add a Kafka dependency to the `pom.xml` project:

```xml
<!-- Kafka -->
    <dependency>
        <groupId>io.micronaut.kafka</groupId>
        <artifactId>micronaut-kafka</artifactId>
    </dependency>
...
```

   By importing the `micronaut-kafka` dependency, we can leverage the Kafka toolkit in the `pet-clinic-reviews` microservice.

2. Once the dependency has been imported, we will then need to configure `application.properties` as follows:

```
micronaut:
  application:
    name: PetClinicReviews
  server:
    port: 8083
kafka:
  bootstrap:
    servers: localhost:9094
```

As mentioned in preceding `application.properties`, we will fix port `8083` for the `pet-clinic-reviews` microservice and configure the Kafka connection by providing Bootstrap server details.

3.  Next, we will create a Kafka client in the `pet-clinic-reviews` microservice, which can send messages to the `vet-reviews` topic. Begin by creating a package, `com.packtpub.micronaut.integration.client`. This package will contain the required client and, in the future, may contain more artifacts related to service integration. We now add `VetReviewClient` to this package:

```
@KafkaClient
public interface VetReviewClient {
    @Topic("vet-reviews")
    void send(@Body VetReviewDTO vetReview);
}
```

`VetReviewClient` is annotated with `@KafkaClient`. Using the `@KafkaClient` annotation, we can inject `VetReviewClient` as a Kafka client. Furthermore, just by simply using `@Topic("vet-reviews")`, we can send the messages (no need to even create the topic) to the `vet-reviews` topic stream.

Hitherto, we have configured application properties and created a simple Kafka client. In the following code, we will make changes to `createVetReview()` in `VetReviewResource` so it can send messages to the topic stream when a new vet review is posted:

```
@Post("/vet-reviews")
@ExecuteOn(TaskExecutors.IO)
public HttpResponse<VetReviewDTO> createVetReview(@Body
VetReviewDTO vetReviewDTO) throws URISyntaxException {
    log.debug("REST request to save VetReview : {}",
vetReviewDTO);
    if (vetReviewDTO.getReviewId() != null) {
        throw new BadRequestAlertException("A new vetReview
cannot already have an ID", ENTITY_NAME, "idexists");
    }
    VetReviewDTO result = vetReviewService.save(vetReviewDTO);

    /** Stream to other services */
    vetReviewClient.send(result);
```

```
    URI location = new URI("/api/vet-reviews/" + result.
getReviewId());
    return HttpResponse.created(result).headers(headers -> {
        headers.location(location);
        HeaderUtil.createEntityCreationAlert(headers,
applicationName, true, ENTITY_NAME, result.getReviewId());
    });
}
```

From the preceding code, we can see that we can simply inject `VetReviewClient` into `VetReviewResource`. In `createVetReview()`, when a vet review is successfully inserted, we can send the message to the `vet-reviews` stream using `VetReviewClient`.

In this section, we introduced the event producer in the `pet-clinic-reviews` microservice. In the following section, we will verify this event producer by invoking the HTTP POST endpoint to create a new vet review.

## Testing the event producer in the pet-clinic-reviews microservice

To test the event producer that has just been created, boot up the `pet-clinic-reviews` microservice locally and access the HTTP POST endpoint. In the following screenshot, we are using a REST client to invoke the `vet-reviews` HTTP POST endpoint to create a vet review:

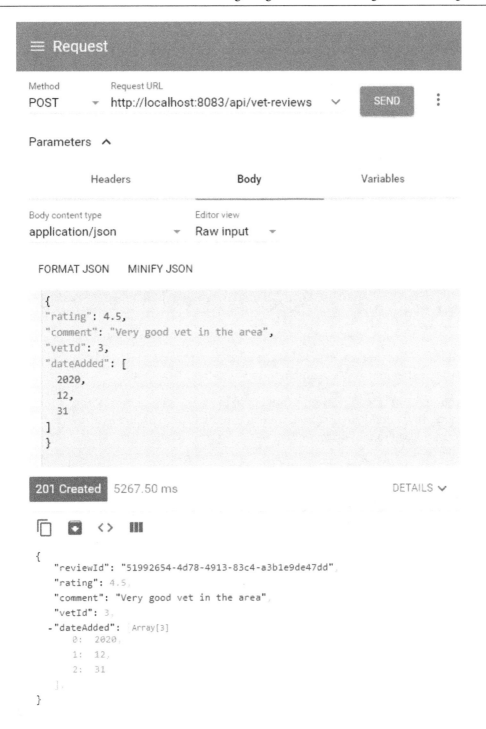

Figure 5.7 – Creating a vet review for testing the event producer

As shown in the preceding screenshot, when we submit a request to create a new vet review, it will persist the vet review and also stream out the review to Apache Kafka. This event message can be verified by accessing Kafdrop at `http://localhost:9100/`. This is what the screen outputs:

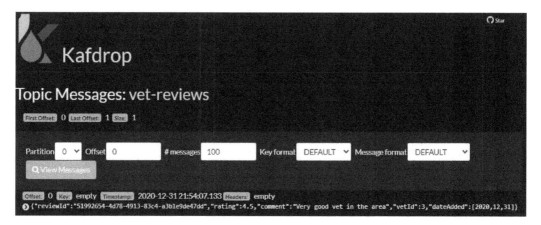

Figure 5.8 – A newly added review to the vet-reviews stream

As viewed in Kafdrop, we can verify that the event from the `pet-clinic-reviews` microservice is streamed out and added to the `vet-reviews` topic.

In this section, we verified the event producer in the `pet-clinic-reviews` microservice. In the following section, we will explore how to implement an event consumer in the Micronaut framework.

## Implementing an event consumer client in the pet-clinic microservice

In this section, we will implement an event consumer in the `pet-clinic` microservice so that it can consume messages streamed in the `vet-reviews` topic.

To begin, we will need to add a Kafka dependency to the pom.xml project. This is shown with the following code:

```
<!-- Kafka -->
    <dependency>
        <groupId>io.micronaut.kafka</groupId>
        <artifactId>micronaut-kafka</artifactId>
    </dependency>
...
```

Importing micronaut-kafka will enable us to leverage the Kafka consumer toolkit. Once the dependency has been imported, we will then need to configure application. properties as follows:

```
micronaut:
  application:
    name: Pet-Clinic
  server:
    port: 8082
kafka:
  bootstrap:
    servers: localhost:9094
```

As mentioned in the preceding code, we will fix port 8082 for the pet-clinic microservice and configure the Kafka connection by providing Bootstrap server details.

Next, to contain all the Kafka integration artifacts, we will create a com.packtpub. micronaut.integration package. Since we will be consuming from the vet-reviews topic stream, we will add VetReviewDTO to the com.packtpub. micronaut.integration.domain package.

Some developers advocate keeping the DTOs in a shared repository that can be re-used in all microservices. However, keeping all the DTOs under an owning microservice is good for better encapsulation. Furthermore, there could be cases where a DTO such as VetReviewDTO could assume the desired object definition in one microservice and a different one in another microservice.

We will create a Kafka listener in the `com.packtpub.micronaut.integration.` `client` package to leverage the `micronaut-kafka` toolkit. Refer to the following code block:

```
@KafkaListener(groupId = "pet-clinic")
public class VetReviewListener {
    private static final Logger log =  LoggerFactory.
getLogger(VetReviewListener.class);
    private final VetService vetService;
    public VetReviewListener(VetService vetService) {
        this.vetService = vetService;
    }
    @Topic("vet-reviews")
    public void receive(@Body VetReviewDTO vetReview) {
        log.info("Received: vetReview -> {}", vetReview);
        try {
            vetService.updateVetAverageRating(vetReview.
getVetId(), vetReview.getRating());
        } catch (Exception e) {
            log.error("Exception occurred: {}", e.toString());
        }
    }
}
```

From the preceding code, we see that we created the `VetReviewListener` using the `@KafkaListener` annotation. In the `@KafkaListener` annotation, we passed `groupId`. Assigning a group ID to a Kafka listener adds the listener to a consumer group. This may be required when there are multiple consumer services for a topic stream so that the Kafka ecosystem can maintain an isolated offset for each consumer. Using `@Topic("vet-reviews")` allows `VetReviewListener` to receive any streamed out messages from the `vet-reviews` stream. When `VetReviewListener` receives any message, it invokes `updateVetAverageRating()` in `VetService`. In the following code snippet, we added this method in `VetService` to update the average rating for a vet when a new review is added to the `pet-clinic-reviews` microservice:

```
public void updateVetAverageRating(Long id, Double rating)
throws Exception {
    log.debug("Request to update vet rating, id: {}, rating:
{}", id, rating);
```

```
    Optional<VetDTO> oVetDTO = findOne(id);
   if (oVetDTO.isPresent()) {
        VetDTO vetDTO = oVetDTO.get();

        Double averageRating = vetDTO.getAverageRating() !=
null ? vetDTO.getAverageRating() : 0D;
        Long ratingCount = vetDTO.getRatingCount() != null ?
vetDTO.getRatingCount() : 0L;
        Double newAvgRating = ((averageRating * ratingCount) +
rating) / (ratingCount + 1);
        Long newRatingCount = ratingCount + 1;

        vetRepository.updateVetAverageRating(id, newAvgRating,
newRatingCount);
    }
}
```

From the preceding code, we see that the updateVetAverageRating() method retrieves the last stored rating. If the last stored rating is null, it assumes it to be 0. In any case, it will add on the new rating and determine a new average rating. Once the average rating has been determined, rating information is persisted in the database by making a call to the repository.

In this section, we explored how we can implement an event consumer in the pet-clinic microservice. In the following section, we will verify this event consumer by creating a new vet review.

## Testing the event consumer in the pet-clinic microservice

To test the event consumer that has just been created, we can boot up the `pet-clinic` (event consumer) and `pet-clinic-reviews` (event producer) microservices. Once the `pet-clinic-reviews` microservice is running, add a new vet review. In the following screenshot, you can see that we are using an HTTP REST client to post a vet review:

Figure 5.9 – Adding a new vet review to test the event consumer

In the POST request to the vet-reviews resource, we are adding an abysmal rating. The pet-clinic-reviews microservice successfully executed the request and responded with an HTTP 201 response, assigning a review ID to the review submitted.

As shown in the following screenshot, in the pet-clinic microservice, if we put a debug point in VetReviewListener, we can verify that the Kafka topic stream is sending out the message for a new vet review:

Figure 5.10 – Event message received by the event consumer

As shown in the preceding screenshot, when the pet-clinic-reviews microservice produces an event message, it is received by the pet-clinic microservice. This is the magic that integrates these two microservices using event-driven architecture. And this pattern can be extended to integrate microservices in a variety of different scenarios, such as a service sending out a message to multiple microservices or chained event messages, or choreographing complex microservice integrations.

In this section, we verified the event consumer in the pet-clinic microservice such that when a new vet review is added to pet-clinic-reviews, pet-clinic receives the review information from the vet-reviews topic stream.

# Summary

In this chapter, we started things off with some fundamentals of event-driven architecture, discussing two different kinds of event publishing models, which are pub/sub and event streaming. We discussed the core components of each model, as well as the pros/cons of using each model.

Since event streaming was better suited for the pet-clinic application, we dived into event streaming using the Apache Kafka ecosystem. For hands-on exercises, we integrated the `pet-clinic-reviews` and the `pet-clinic` microservices using an Apache Kafka topic stream. We verified the integration by creating a new vet review and received the rating in the `pet-clinic` microservice to update the average rating for a vet.

This chapter has provided you with a solid understanding of event-driven architecture and a practical skillset in implementing an event-streaming system in the Micronaut framework.

In the next chapter, we will explore how we can automate quality testing using built-in as well as third-party tools in the Micronaut framework.

# Questions

1. What is event-driven architecture?
2. What is the pub/sub model in event-driven architecture?
3. What is event streaming?
4. Describe the various components that make up the Apache Kafka ecosystem.
5. How is the Apache ecosystem set up in Docker?
6. How are microservices integrated in the Micronaut framework using event streaming?
7. How is an event consumer implemented in the Micronaut framework?
8. How is an event producer implemented in the Micronaut framework?

# Section 3: Microservices Testing

This section will focus on microservices testing in the Micronaut framework and has the following chapter:

- *Chapter 6, Testing the Microservices*

# 6
# Testing
# Microservices

In a rather simple definition, **software testing** is verifying that a produced software application is functioning as expected. Since the early days of programming languages and software development, good precedents have been set to ensure they are **functioning as expected**. Almost all programming languages (barring some scripting languages) have robust compilers to catch anomalies at compile time. Though compile-time checks are good to start with, they can't verify whether a software application will run just as expected at runtime. For peace of mind, software development teams perform various kinds of testing to verify that a software application will function as expected. And any testing exercise will increase manifold with an increase in the number of distributed components or, put simply, it's rather more easy to test a monolithic application than a distributed one. To save time and decrease the turnaround time to deliver a feature, it's efficient to automate testing at various levels.

In this chapter, we will explore how we can automate testing at various levels of microservices. We will dive into the following topics:

- Understanding the testing pyramid
- Unit testing in the Micronaut framework
- Service testing in the Micronaut framework
- Integration testing using test containers

By the end of this chapter, you will have handy knowledge of automating testing at various levels of microservices.

# Technical requirements

All the commands and technical instructions in this chapter are run on Windows 10 and macOS. Code examples covered in this chapter are available on the book's GitHub at `https://github.com/PacktPublishing/Building-Microservices-with-Micronaut/tree/master/Chapter06`.

The following tools need to be installed and set up in the development environment:

- **Java SDK** version 13 or above (we used Java 14)

- **Maven**: It is optional and only required if you would like to use Maven as the build system. However, we recommend having Maven set up on any development machine. Instructions to download and install Maven can be found at `https://maven.apache.org/download.cgi`.

- **Development IDE**: Based on your preference, any Java-based IDE can be used, but for the purpose of writing this chapter, IntelliJ was used.

- **Git**: Instructions to download and install it can be found at `https://git-scm.com/downloads`.

- **PostgreSQL**: Instructions to download and install it can be found at `https://www.postgresql.org/download/`.

- **MongoDB**: MongoDB Atlas provides a free online database-as-a-service offering of up to 512 MB storage. However, if a local database is preferred, then instructions to download and install it can be found at `https://docs.mongodb.com/manual/administration/install-community/`. We used a local installation to write this chapter.

- **REST client**: Any HTTP REST client can be used. We used the Advanced REST Client Chrome plugin.

- **Docker**: Instructions to download and install Docker can be found at `https://docs.docker.com/get-docker/`.

# Understanding the testing pyramid

The testing pyramid is an easy concept to understand the relative notion of performance, expense, and robustness for the different kinds of testing. The following diagram shows various kinds of testing in the testing pyramid and how much effort is required:

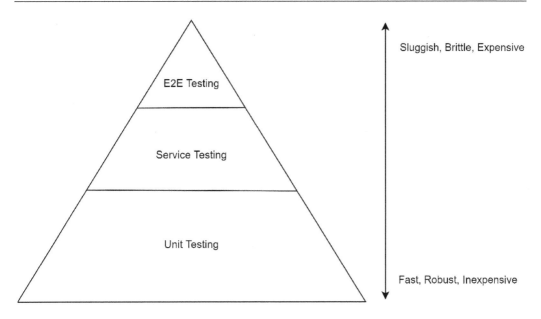

Figure 6.1 – Test pyramid

As depicted in the preceding diagram, unit testing is fast, robust, and inexpensive whereas as we go towards the top of the pyramid, testing becomes sluggish, brittle, and expensive. Though all kinds of testing are required to fully verify whether the application is working as expected, a fine balance is critical to cut expenses and increase robustness and speed. Put simply, have a lot of unit tests, a number of service tests, and very few end-to-end tests. This will ensure quality at a greater speed and with less cost.

In the next section, we will begin the automated testing journey with unit testing.

## Unit testing in the Micronaut framework

In the object-oriented paradigm, an object can assume multiple behaviors. These behaviors are defined by their methods. Effective unit testing probes a single behavior of an object at a time. This doesn't translate to testing a method as a method can change its behavior by taking different execution paths (if the method has forked control flow). Therefore, essentially, a unit test will probe one execution path of a method at a time. Iteratively, we can add more unit tests to probe other execution paths in the same or other methods. This bottom-up approach relies on verifying behaviors at smaller, isolated levels so the application as a whole will work as expected.

To perform a unit test, isolation is required. Essentially, we need to isolate the object's behavior (that we want to test) while ignoring the object's interaction with other objects/components within the system. To achieve this isolation, we have various mechanisms in unit testing:

- **Mocking**: Mocking is an operation to create a test double in which a testing framework will create a mock/dummy object based on the class definition of the object (compile time). To isolate the subject object's interaction with other interacting objects, we can simply mock the interacting objects. When a unit test is executed for the subject object, it will skip interactions with other objects.

- **Spying**: Using spying we create a test double by probing the actual instance of the object (runtime). A spy object will just be the same as the real object barring any stubs. Stubs are used to define a dummy invocation so that the spy object will execute normally but when an invocation matches a stub definition, then it will execute the dummy behavior defined by the stub.

Though mocking and spying can help to isolate the behavior, sometimes the subject object may not be interacting with other objects so no test doubles are required. In the next section, we will begin with how to implement unit tests in the Micronaut framework using JUnit 5.

## Unit testing using JUnit 5

In order to learn how to implement unit tests in the Micronaut framework, we will resume the code base from *Chapter 5, Integrating Microservices Using the Event-Driven Architecture*. We will continue with the pet-owner microservice and make sure you have the following dependencies added to the project pom.xml:

```xml
<dependency>
  <groupId>org.junit.jupiter</groupId>
  <artifactId>junit-jupiter-api</artifactId>
  <scope>test</scope>
</dependency>
<dependency>
  <groupId>org.junit.jupiter</groupId>
  <artifactId>junit-jupiter-engine</artifactId>
  <scope>test</scope>
</dependency>
<dependency>
  <groupId>io.micronaut.test</groupId>
```

```
            <artifactId>micronaut-test-junit5</artifactId>
            <scope>test</scope>
        </dependency>
```

By importing the preceding JUnit dependencies, we can leverage the JUnit and Micronaut test toolkit in the pet-owner microservice.

Next, we will create a `TestUtil` class in `com.packtpub.micronaut.util` that can encapsulate some essential testing methods:

```
public final class TestUtil {
    public static <T> void equalsVerifier(Class<T> clazz)
        throws Exception {
            T domainObject1 =
              clazz.getConstructor().newInstance();
            assertThat(domainObject1.toString()).isNotNull();
            assertThat(domainObject1).isEqualTo(domainObject1);
            assertThat(domainObject1.hashCode()).
    isEqualTo(domainObject1.hashCode());
            // Test with an instance of another class
            Object testOtherObject = new Object();
            assertThat(domainObject1).
    isNotEqualTo(testOtherObject);
            assertThat(domainObject1).isNotEqualTo(null);
            // Test with an instance of the same class
            T domainObject2 =
              clazz.getConstructor().newInstance();
            assertThat(domainObject1).isNotEqualTo(domainObject2);
            /* HashCodes are equals because the objects are not
    persisted yet */
            assertThat(domainObject1.hashCode()).
    isEqualTo(domainObject2.hashCode());
        }
}
```

In `TestUtil`, we have added the `equalsVerifier()` method, which can verify whether two objects are equal or not. This method takes a class type as an input parameter to assert different conditions on tested objects.

In the next section, we will explore how to unit test a domain object.

## Unit testing a domain object

A domain object is simply a **POJO** (**plain old Java object**) and we can create a basic test for an `Owner` class in the pet-owner microservice. In the followed code snippet, we are creating an `OwnerTest` class to assert the equality of two owner instances:

```java
public class OwnerTest {
    @Test
    public void equalsVerifier() throws Exception {
        TestUtil.equalsVerifier(Owner.class);
        Owner owner1 = new Owner();
        owner1.setId(1L);
        Owner owner2 = new Owner();
        owner2.setId(owner1.getId());
        assertThat(owner1).isEqualTo(owner2);
        owner2.setId(2L);
        assertThat(owner1).isNotEqualTo(owner2);
        owner1.setId(null);
        assertThat(owner1).isNotEqualTo(owner2);
    }
}
```

The `OnwerTest` class contains a test method, `equalsVerifier()`. An annotation, `org.junit.jupiter.api.Test`, is used to mark it as a test method. To verify the expected behavior, we are using assert statements. Similarly, we can define test classes for other domain objects in the pet-owner microservice.

In the next section, we will unit test a mapper object.

## Unit testing a mapper object

Our mapper objects in the pet-owner microservice are simple and we can create a basic test for an `OwnerMapper` class using the `@Test` annotation. In the following code snippet, `OwnerMapperTest` is unit-testing the `fromId()` method in `OwnerMapper`:

```java
public class OwnerMapperTest {

    private OwnerMapper;
```

```
@BeforeEach
public void setUp() {
    ownerMapper = new OwnerMapperImpl();
}


@Test
public void testEntityFromId() {
    Long id = 1L;
    assertThat(ownerMapper.fromId(id).getId().
      isEqualTo(id);
    assertThat(ownerMapper.fromId(null)).isNull();
}
}
```

The `OwnerMapperTest` class contains a test method, `testEntityFromId()`. To verify the expected behavior we are using `assert` statements. Similarly, we can define test classes for other mapper objects in the pet-owner microservice.

Hitherto, we wrote simple unit tests for domain and mapper objects that didn't require any test doubles. In the next section, we will explore how we can use mocking to create the desired test doubles.

## Using mocks in unit testing

As we discussed before, mocking a testing framework will create a test double based on the class definition. These test doubles come in handy in unit testing an object where the object invokes methods on other objects.

In order to learn about mocking in unit testing, we will work on the `VetService` class in the pet-clinic microservice. Let's look at `VetServiceImpl` in the pet-clinic microservice:

```
@Singleton
public class VetServiceImpl implements VetService {
    private final VetRepository;
    private final SpecialtyRepository;
    private final VetMapper;
    private final SpecialtyMapper;
    public VetServiceImpl(VetRepository,
      SpecialtyRepository, VetMapper, SpecialtyMapper
```

```
    specialtyMapper) {
        this.vetRepository = vetRepository;
        this.specialtyRepository = specialtyRepository;
        this.vetMapper = vetMapper;
        this.specialtyMapper = specialtyMapper;
    }
    ...
}
```

VetService instantiates VetRepository, SpecialtyRepository, VetMapper, and SpecialtyMapper in the constructor. These instantiated objects are then used in VetService methods. To unit test the VetService object, we would need to define mocks for some of these interacting objects.

Let's create VetServiceTest to encapsulate unit tests for VetService. In this test class, we will mock some interacting objects using the @MockBean annotation:

```
@MicronautTest
class VetServiceTest {
    @Inject
    private VetRepository;
    @Inject
    private SpecialtyRepository;
    @Inject
    private VetMapper;
    @Inject
    private SpecialtyMapper;
    @Inject
    private VetService;

    /** Mock beans */
    @MockBean(VetRepositoryImpl.class)
    VetRepository vetRepository() {
        return mock(VetRepository.class);
    }
    @MockBean(SpecialtyRepositoryImpl.class)
    SpecialtyRepository specialtyRepository() {
        return mock(SpecialtyRepository.class);
```

```
        }

        ...

}
```

The `VetServiceTest` class is annotated with the `@MicronautTest` annotation. It runs the test class as an actual Micronaut application with the full application context, thereby avoiding the artificial separation between production code and test code.

To inject the interacting objects, we are using `@Inject` annotations. `@Inject` injects a bean from the application context into the class. Furthermore, using the `@MockBean` annotation, we are overriding the runtime beans for `VetRepository` and `SpecialtyRepository`. `@MockBean` will replace the actual objects with mocked objects in the application context.

We can easily use these test double mocks in writing a unit test for the `VetService` method:

```
@Test
public void saveVet() throws Exception {
    // Setup Specialty
    Long specialtyId = 100L;
    SpecialtyDTO = createSpecialtyDTO(specialtyId);
    Specialty = specialtyMapper.toEntity(specialtyDTO);

    // Setup VetDTO
    Long vetId = 200L;
    VetDTO = createVetDTO(vetId);
    vetDTO.setSpecialties(Set.of(specialtyDTO));
    Vet = vetMapper.toEntity(vetDTO);

    // Stubbing
    when(vetRepository.save(any(Vet.class))).thenReturn
      (vetId);
    when(specialtyRepository.findByName(anyString())).
      thenReturn(specialty);
    doNothing().when(vetRepository).saveVetSpecialty
      (anyLong(), anyLong());
    when(vetRepository.findById(anyLong())).thenReturn
```

```
    (vet);

    // Execution
    VetDTO savedVetDTO = vetService.save(vetDTO);

    verify(vetRepository, times(1)).save(any(Vet.class));
    verify(specialtyRepository,
      times(1)).findByName(anyString());
    verify(vetRepository, times(1)).saveVetSpecialty
      (anyLong(), anyLong());
    verify(vetRepository, times(1)).findById(anyLong());

    assertThat(savedVetDTO).isNotNull();
    assertThat(savedVetDTO.getId()).isEqualTo(vetId);
    assertThat(savedVetDTO.getSpecialties()).isNotEmpty();
    assertThat(savedVetDTO.getSpecialties().size()).
      isEqualTo(1);
    assertThat(savedVetDTO.getSpecialties().stream().
      findFirst().orElse(null).getId()).isEqualTo
      (specialtyId);
}
```

In the preceding code snippet, you can see how we are defining stubs for the mocked
VetRepository and SpecialtyRepository classes. Usually, a mock stub takes the
form of when(object.methodCall()).thenReturn(result), except in the case
of void method calls, where it is doNothing().when(object).methodCall().

Ideally, it's prudent to follow up mock stubs with verify() statements. verify() will
confirm that, indeed, the desired method calls were made while executing the unit test.

In the next section, we will explore another way of creating test doubles using spies.

## Using spies in unit testing

As we discussed before, spying on a testing framework will create a test double based on
the actual runtime object of the class. While mocking creates a full test double of the real
object, with spying we can control whether the test double is partial or full. In a spied
object, we can stub some method calls while keeping other method calls real. In such
a scenario, the unit test will make dummy as well as real calls. Therefore, spying gives
a bit more control over what we want to fake.

In order to learn about spying in unit testing, we will work on the `SpecialtyService` class in the pet-clinic microservice. Let's look at `SpecialtyServiceImpl` in the pet-clinic microservice:

```
public class SpecialtyServiceImpl implements SpecialtyService {
    private final SpecialtyRepository;
    private final SpecialtyMapper;
    public SpecialtyServiceImpl(SpecialtyRepository
    specialtyRepository, SpecialtyMapper specialtyMapper) {
        this.specialtyRepository = specialtyRepository;
        this.specialtyMapper = specialtyMapper;
    }
    ...
}
```

`SpecialtyService` is instantiating `SpecialtyRepository` and `SpecialtyMapper` in the constructor. These instantiated objects are then used in `SpecialtyService` methods. To unit test the `SpecialtyService` object, we would need to define spies for some of these interacting objects.

Let's create `SpecialtyServiceTest` for encapsulating unit tests for `SpecialtyService`. In this test class, we will spy some interacting objects using the @MockBean annotation and `spy()` method in JUnit:

```
@MicronautTest
class SpecialtyServiceTest {
    @Inject
    private SpecialtyRepository;
    @Inject
    private SpecialtyMapper;
    @Inject
    private SpecialtyService;
    @MockBean(SpecialtyRepositoryImpl.class)
    SpecialtyRepository specialtyRepository() {
        return spy(SpecialtyRepository.class);
    }
    ...
}
```

The `SpecialtyServiceTest` class is annotated with `@MicronautTest`, which runs the test class as an actual Micronaut application with the full application context.

Using the `@MockBean` annotation, we are overriding the runtime bean for `SpecialtyRepository`. `@MockBean` will replace the actual object with the spied object in the application context. On the spied `SpecialtyRepository` object, we can easily define some stubs that will be executed in the test method instead of the actual invocation:

```
@Test
public void saveSpecialty() throws Exception {
    // Setup Specialty
    Long specialtyId = 100L;
    SpecialtyDTO = createSpecialtyDTO(specialtyId);
    Specialty = specialtyMapper.toEntity(specialtyDTO);

    // Stubbing
    doReturn(100L).when(specialtyRepository).save(any
      (Specialty.class));
    doReturn(specialty).when(specialtyRepository).findById
      (anyLong());

    // Execution
    SpecialtyDTO savedSpecialtyDTO =
        specialtyService.save(specialtyDTO);

    verify(specialtyRepository,
      times(1)).save(any(Specialty.class));
    verify(specialtyRepository,
      times(1)).findById(anyLong());

    assertThat(savedSpecialtyDTO).isNotNull();
    assertThat(savedSpecialtyDTO.getId()).isEqualTo
        (specialtyId);
}
```

In the preceding code snippet, you can see how we are defining stubs for the spied `SpecialtyRepository` instance. Usually, a spy stub takes the form of `doReturn(result).when(object).methodCall()`, except in the case of void method calls, where it is `doNothing().when(object).methodCall()`.

Again, it's prudent to follow up spied stubs with `verify()` statements. These will confirm whether the desired method calls were made while executing the unit test.

Hitherto, we have learned the various ways to unit test using mocks and spies. In the next section, we will explore how we can perform service testing in the Micronaut framework.

# Service testing in the Micronaut framework

**Service testing** is the next level to unit testing. By testing all the endpoints in a microservice and repeating this process for all the other microservices, we can make sure that all the services are working as expected edge to edge. It raises the quality check to the next level. Having said that, as we discussed before, as we move up in the test pyramid, test cases become more brittle, expensive, and sluggish, therefore, we need to establish a fine balance of not testing too much on the higher levels.

To learn how we can perform service testing in the Micronaut framework, we will continue with the pet-clinic microservice. In the following sections, we will go into testing all the REST endpoints of a service. We will use the `@Order` annotation to establish the order of execution of a test in the suite. An ordered test suite can help in starting from scratch and cleaning up at the end. In the following examples, we will create, get, update, and finally delete the resource.

## Creating the test suite

To test the `VetResource` endpoints, let's create a `VetResourceIntegrationTest` class. This suite will encapsulate all the happy and unhappy integration tests:

```java
@MicronautTest(transactional = false)
@Property(name = "micronaut.security.enabled", value = "false")
@TestInstance(TestInstance.Lifecycle.PER_CLASS)
@TestMethodOrder(MethodOrderer.OrderAnnotation.class)
public class VetResourceIntegrationTest {
    @Inject
    private VetMapper;
    @Inject
    private VetRepository;
```

```
@Inject
private SpecialtyRepository;
@Inject @Client("/")
RxHttpClient client;

...
}
```

There are a few things to ponder in the preceding code snippet:

- **@MicronautTest(transactional = false)**: This annotation boots up the test suite as a real Micronaut application and `transactional = false` ensures that the suite runs without the transaction.

- **@Property(name = "micronaut.security.enabled", value = "false")**: The `@Property` annotation overrides the application configuration. And in our case, we are disabling the security.

- **@TestInstance(TestInstance.Lifecycle.PER_CLASS)**: `TestInstance.Lifecycle.PER_CLASS` boots up the instance and keeps the application context for the whole suite. You can instantiate a test object and application context at the test method level using `@TestInstance(TestInstance.Lifecycle.PER_METHOD)`.

- **@TestMethodOrder(MethodOrderer.OrderAnnotation.class)**: `@TestMethodOrder annotation` in JUnit is used to define the execution order of each test method in the test suite.

- **@Inject @Client**: This annotation injects a reactive HTTP client (built in Micronaut) to perform RESTful calls to the resource endpoints.

After setting up the test suite, we are good to perform service testing. In the next few sections, we will cover all the restful calls in the test methods.

## Testing the create endpoint

`VetResource` has a POST endpoint for creating a new `Vet`. It accepts `VetDTO` in the request body. Let's use the HTTP client to create a vet:

```
@Test
@Order(1)
public void createVet() throws Exception {
    int databaseSizeBeforeCreate =
      vetRepository.findAll().size();
```

```java
        VetDTO = vetMapper.toDto(vet);

        // Create the Vet
        HttpResponse<VetDTO> response =
         client.exchange(HttpRequest.POST("/api/vets", vetDTO),
         VetDTO.class).blockingFirst();
        assertThat(response.status().getCode()).isEqualTo
          (HttpStatus.CREATED.getCode());

        // Validate the Vet in the database
        List<Vet> vetList = (List<Vet>)
          vetRepository.findAll();
        assertThat(vetList).hasSize(databaseSizeBeforeCreate +
         1);
        Vet testVet = vetList.get(vetList.size() - 1);

        // Set id for further tests
        vet.setId(testVet.getId());

        assertThat(testVet.getFirstName()).isEqualTo
         (DEFAULT_FIRST_NAME);
        assertThat(testVet.getLastName()).isEqualTo
         (DEFAULT_LAST_NAME);
}
```

In the preceding test, we are creating a VetDTO object and invoking the POST endpoint using the HTTP client. To make the reactive client return the observable and make a pseudo-synchronous call, we are using blockingFirst(). It blocks the thread until the observable emits an item, then returns the first item emitted by the observable. Finally, we are asserting to confirm the expected versus the actual behavior.

## Testing the GET endpoint

In the proceeding test, we created a new vet in the POST endpoint service test. We can leverage just the persisted vet to test the GET endpoint:

```java
@Test
@Order(3)
public void getAllVets() throws Exception {
    // Get the vetList w/ all the vets
    List<VetDTO> vets = client.retrieve(HttpRequest.GET
      ("/api/vets?eagerload=true"),
        Argument.listOf(VetDTO.class)).blockingFirst();
    VetDTO testVet = vets.get(vets.size() - 1);

    assertThat(testVet.getFirstName()).isEqualTo
      (DEFAULT_FIRST_NAME);
    assertThat(testVet.getLastName()).isEqualTo
      (DEFAULT_LAST_NAME);
}


@Test
@Order(4)
public void getVet() throws Exception {
    // Get the vet
    VetDTO testVet =
        client.retrieve(HttpRequest.GET("/api/vets/" +
        vet.getId()), VetDTO.class).blockingFirst();

    assertThat(testVet.getFirstName()).isEqualTo
      (DEFAULT_FIRST_NAME);
    assertThat(testVet.getLastName()).isEqualTo
      (DEFAULT_LAST_NAME);
}
```

In the preceding tests, we are testing two endpoints, getVet() and getAllVets(). To make the reactive client return the results, we are using the blockingFirst() operator. While getAllVets() will return a list of vets, getVet() will return the desired vet object only.

## Testing the update endpoint

To test the `update` endpoint, we will leverage the vet resource persisted in the created endpoint service test, therefore, use an order after the `create` and `GET` calls:

```
@Test
@Order(6)
public void updateVet() throws Exception {
    int databaseSizeBeforeUpdate =
      vetRepository.findAll().size();


    // Update the vet
    Vet updatedVet = vetRepository.findById(vet.getId());


    updatedVet
        .firstName(UPDATED_FIRST_NAME)
        .lastName(UPDATED_LAST_NAME);
    VetDTO updatedVetDTO = vetMapper.toDto(updatedVet);


    @SuppressWarnings("unchecked")
    HttpResponse<VetDTO> response =
      client.exchange(HttpRequest.PUT("/api/vets",
        updatedVetDTO), VetDTO.class)
          .onErrorReturn(t -> (HttpResponse<VetDTO>)
        ((HttpClientResponseException)
          t).getResponse()).blockingFirst();


    assertThat(response.status().getCode()).isEqualTo
      (HttpStatus.OK.getCode());


    // Validate the Vet in the database
    List<Vet> vetList = (List<Vet>)
      vetRepository.findAll();
    assertThat(vetList).hasSize(databaseSizeBeforeUpdate);
    Vet testVet = vetList.get(vetList.size() - 1);


    assertThat(testVet.getFirstName()).isEqualTo
      (UPDATED_FIRST_NAME);
```

```
    assertThat(testVet.getLastName()).isEqualTo
        (UPDATED_LAST_NAME);
}
```

In the preceding test, we tested the updateVet() endpoint. We first fetched the persisted vet and then updated the first and last name before invoking the update endpoint. Finally, we asserted to confirm the actual behavior meets the expected behavior.

## Testing the delete endpoint

To test the delete endpoint, we will leverage the vet resource persisted in the earlier endpoint calls. Therefore, we will use an order after the create, GET, and update calls:

```
@Test
@Order(8)
public void deleteVet() throws Exception {
    int databaseSizeBeforeDelete =
     vetRepository.findAll().size();

    // Delete the vet
    @SuppressWarnings("unchecked")
    HttpResponse<VetDTO> response =
     client.exchange(HttpRequest.DELETE("/api/vets/"+
     vet.getId()), VetDTO.class)
        .onErrorReturn(t -> (HttpResponse<VetDTO>)
        ((HttpClientResponseException)
        t).getResponse()).blockingFirst();

    assertThat(response.status().getCode()).isEqualTo
     (HttpStatus.NO_CONTENT.getCode());

    // Validate the database is now empty
    List<Vet> vetList = (List<Vet>)
     vetRepository.findAll();
    assertThat(vetList).hasSize
        (databaseSizeBeforeDelete - 1);
}
```

In the preceding test, we tested the `deleteVet()` endpoint. We are passing the previously persisted `vetId`. And after the successful service call, we are asserting to confirm the actual behavior meets the expected behavior by comparing the database size before and after the service call.

The test orders in the suite ensure that we always start from scratch and leave it clean after finishing all the tests in the suite. There are pros and cons to this pattern for service testing compared to setting up and cleaning up at the test method level. You can pick and choose a pattern after analyzing the application requirements and whether to use suite setup and cleanup or at the test method level.

In the next section, we will explore the exciting world of `Testcontainers` for integration testing.

# Integration testing using Testcontainers

`Testcontainers` is a Java library that elegantly marries the world of testing with Docker virtualization. Using the `Testcontainers` library, we can set up, instantiate, and inject any Docker container into the testing code. This approach opens up many avenues for performing integration testing. In the test suite or test method setup, we can boot up a Dockerized database, Kafka or email server or any integrating app, perform the integration tests, and destroy the Dockerized app in the cleanup. With this pattern, we are up close to the production environment while not impacting the environment with any after-testing side effects.

To learn how we can use the `Testcontainers` library, we will experiment on the pet-clinic-reviews microservice that integrates with MongoDB. In the next section, we will begin setting up `Testcontainers` in the Micronaut application.

## Setting up the Testcontainers in the Micronaut application

To use `Testcontainers` in the pet-clinic-reviews microservice, add the following dependencies in the project `pom.xml`:

```xml
<!-- Test containers -->
<dependency>
    <groupId>org.testcontainers</groupId>
    <artifactId>junit-jupiter</artifactId>
    <version>1.15.2</version>
    <scope>test</scope>
```

```
    </dependency>
    <dependency>
      <groupId>org.testcontainers</groupId>
      <artifactId>mongodb</artifactId>
      <version>1.15.2</version>
      <scope>test</scope>
    </dependency>
```

By importing a MongoDB flavored test container, we will be able to leverage the MongoDB Docker toolkit. After importing the required `Testcontainers` dependencies, let's set up an abstract class that can provide any app containers required by the integration tests:

```
public class AbstractContainerBaseTest {
    public static final MongoDBContainer
     MONGO_DB_CONTAINER;
    static {
        MONGO_DB_CONTAINER = new MongoDBContainer
           (DockerImageName.parse("mongo:4.0.10"));
        MONGO_DB_CONTAINER.start();
    }
}
```

In `AbstractContainerBaseTest`, we configure and boot up a MongoDB instance in Docker statically. The static nature of this container will simplify access and avoid booting up too many instances at the test suite or test method level. `Testcontainers` elegantly, and with minimal code, pulls up a MongoDB Docker image, boots it up, and starts it.

In the next section, we will write integration tests using `Testcontainers` for `VetReviewRepository`.

## Writing integration tests using Testcontainers

In the preceding section, we covered how we can use `Testcontainers` to create a Dockerized MongoDB. We will proceed to test `VetReviewRepository` using the Docker MongoDB instance. Let's begin with the test suite and test method setups:

```
@Testcontainers
@TestInstance(TestInstance.Lifecycle.PER_CLASS)
@TestMethodOrder(MethodOrderer.OrderAnnotation.class)
```

```
class VetReviewRepositoryIntegrationTest extends
AbstractContainerBaseTest {
    private VetReviewRepository;
    @BeforeAll
    void init() {
        ApplicationContext context =
        ApplicationContext.run(
            PropertySource.of("test", Map.of
              ("mongodb.uri",
              MONGO_DB_CONTAINER.getReplicaSetUrl()))
        );
        vetReviewRepository =
          context.getBean(VetReviewRepository.class);
    }
    @BeforeEach
    public void initTest() {
        if (!MONGO_DB_CONTAINER.isRunning()) {
            MONGO_DB_CONTAINER.start();
        }
    }
    ...
}
```

In the test suite setup, we are overriding the application properties for MongoDB. Furthermore, we are fetching the VetReviewRepository bean from the application context. This will make sure we are injecting the repository bean that is communicating with the Dockerized MongoDB. And, in the test method setup, we are ensuring that the MongoDB container is running before we execute the test method. Since we have set up at the test suite and test method level, let's jump ahead in writing an integration test:

```
@Test
@Order(1)
public void saveVetReview() {
    VetReview = new VetReview();
    String reviewId = UUID.randomUUID().toString();
    vetReview.setReviewId(reviewId);
    vetReview.setVetId(1L);
    vetReview.setRating(3D);
```

```
        vetReview.setDateAdded(LocalDate.now());
        vetReview.setComment("Good vet");

        vetReviewRepository.save(vetReview);

        VetReview savedVetReview =
            vetReviewRepository.findByReviewId(reviewId);

        assertThat(savedVetReview).isNotNull();
        assertThat(savedVetReview.getReviewId()).isEqualTo
            (reviewId);
    }
```

In the `saveVetReview()` test, we are creating a new vet review and invoking on `VetReviewRepository` to persist this vet review. Finally, we are asserting that the vet review was persisted successfully by fetching and comparing the values. We are using the `@Order` pattern in the test suite so later tests can ensure the cleanup.

In this section, we explored how `Testcontainers` can simplify integration tests by spinning off Docker instances of the database or other service components. We implemented integration tests for `VetReviewRepository` by creating a MongoDB test container.

# Summary

We began this chapter with the testing pyramid and striking a fine balance in test automation with unit testing, service testing, and integration testing. We jumpstarted with some basics of unit testing, such as leveraging mocks and spies to write unit tests. We then dived into how we can write service tests to test various RESTful endpoints using a reactive HTTP client in the Micronaut framework. Finally, we explored the exciting world of test containers for integration testing. We wrote integration tests using `Testcontainer` to instantiate MongoDB in the test environment.

This chapter provides you with a firm understanding of testing at various levels, such as unit, service, or integration in the Micronaut framework. After subtle yet nimble theoretical discussions, we followed up with good hands-on examples to enhance your practical skillset in automated testing in the Micronaut framework.

In the next chapter, we will explore how we can handle microservice architecture concerns in the Micronaut framework.

# Questions

1. What is unit testing?

2. What is mocking in unit testing?

3. How can we mock in the Micronaut framework using JUnit?

4. What is spying in unit testing?

5. How can we spy in the Micronaut framework using JUnit?

6. How can we write service tests in the Micronaut framework?

7. What are test containers?

8. How can you use test containers in the Micronaut framework?

9. How do you write integration tests in the Micronaut framework?

# Section 4: Microservices Deployment

This section will explore the next phase of working on microservices, that is, deployment. Firstly, you will focus on handling some core concerns in microservices, such as service discovery, API gateways, application configurations, and fault tolerance. Later, you will deploy the sample `pet-clinic` application using Docker containers.

This section has the following chapters:

- *Chapter 7, Handling the Microservices Concerns*
- *Chapter 8, Deploying the Microservices*

# 7
# Handling Microservice Concerns

Any microservice architecture implementation is incomplete without handling some fundamental microservice concerns such as **configuration management, API documentation, service discovery, API gateways,** and **fault tolerance**. Hitherto we were focused on the disintegration journey of the microservices, such as how to separate the concerns in modular microservices. For seamless and unified application access, we need the microservices to integrate and expose a coalesced interface. A coalesced interface enables the upstream consumers to interact with backend microservices as though they were one.

A key benefit of implementing microservices is fault tolerance. Fault tolerance mechanisms such as on-demand scaling, fallbacks, and circuit breakers make microservices ubiquitous and robust.

In this chapter, we will explore ways to handle and implement the following microservice concerns:

- Externalizing the application configuration
- Documenting the service APIs

- Implementing service discovery
- Implementing the API gateway
- Implementing the fault-tolerance mechanisms

By the end of this chapter, you will have practical knowledge of handling and implementing these key microservice concerns in the Micronaut framework.

# Technical requirements

All the commands and technical instructions in this chapter run on Windows 10 and macOS. The code examples covered in this chapter are available on the book's GitHub repo at `https://github.com/PacktPublishing/Building-Microservices-with-Micronaut/tree/master/Chapter07`.

The following tools need to be installed and set up in the development environment:

- **Java SDK** version 13 or above (we used Java 14).
- **Maven** – This is optional and only required if you would like to use Maven as the build system. However, we recommend having Maven set up on any development machine. Instructions to download and install Maven can be found at `https://maven.apache.org/download.cgi`.
- **A development IDE** – Based on your preference, any Java-based IDE can be used, but for purpose of writing this chapter, IntelliJ was used.
- **Git** – Instructions to download and install this can be found at `https://git-scm.com/downloads`.
- **PostgreSQL** – Instructions to download and install this can be found at `https://www.postgresql.org/download/`.
- **MongoDB** – MongoDB Atlas provides a free online database-as-a-service with up to 512 MB storage. However, if a local database is preferred then instructions to download and install this can be found at `https://docs.mongodb.com/manual/administration/install-community/`. We used a local installation for writing this chapter.
- **A REST client** – Any HTTP REST client can be used. We used the Advanced REST Client Chrome plugin.
- **Docker** – Instructions to download and install Docker can be found at `https://docs.docker.com/get-docker/`.

# Externalizing the application configuration

Irrespective of the size and complexities of a microservice application, the task of maintaining the configuration settings for each service seems one of the most crucial aspects of working with microservices. Decoupling the service configurations goes back to our earlier discussion on separating the concerns. In the earlier chapters, we have seen how we can handle service configurations using the `application.properties` files. Though this is a step up from not hard-coding these configurations in production code, it's still not enough.

One of the key requirements for any microservice is agility. An ideal microservice should be flexible and rapid in addressing any change in the user's requirements, as well as handling code defects or network issues. Having said that, each enterprise application needs to meet the specific demands of compliance and auditing, which implies that a developer often can't just deploy a code artifact from their workspace directly to a production environment. If the configurations are decoupled from the service code logic then we can easily build the artifact once (without configurations) and deploy it to many environments (where each environment can bootstrap its own configurations).

In the next section, we will dive into how we can manage distributed service configurations in the Micronaut framework.

# Using distributed configuration management to externalize the configuration

Micronaut has out-of-the-box features for **distributed configuration management**. It has built-in support and integration with HashiCorp's **Consul**. Though Consul is a service discovery tool, it also has the capability to store application properties. In the following section, we will use **Consul key-value store** for managing configurations for the `pet-owner` microservice.

### Implementing a configuration store in Consul

We will use a Dockerized Consul instance. Work through the following instructions for installing and running Consul in Docker:

1.  Make sure the Docker application is running in your workspace/environment. Then open a bash terminal (I used Git Bash) and run the following command:

    ```
    docker run -d --name consul -p 8500:8500 consul
    ```

2.  Wait for Docker to download and install **Consul**.

The preceding command will kick-start a single-node Consul instance and expose it on port 8500. We can verify the installation by accessing the Consul web interface at http://localhost:8500/.

Now, to create a configuration store in Consul, follow these instructions:

1. Open the **Consul** web interface at http://localhost:8500/ and select **Key/Value** from the top header.

2. Click on the **Create** button.

3. Type config/pet-owner/application.yml in the **Key or folder** input box.

4. Add the pet-owner application.properties in the text area:

```
micronaut:
  application:
    name: pet-owner
  router:
    static-resources:
      swagger:
        paths: classpath:META-INF/swagger
        mapping: /swagger/**

datasources:
  default:
    url: "jdbc:postgresql://localhost:5432/postgres"
    username: postgres
    password: postgres
    driverClassName: org.postgresql.Driver

jpa:
  default:
    entity-scan:
      packages:
        - com.packtpub.micronaut.domain
    properties:
      hibernate:
        show_sql: false
        dialect: org.hibernate.dialect.
PostgreSQL95Dialect
        enable_lazy_load_no_trans: true
```

By following the preceding steps, we have set up a key-value store in Consul for the pet-owner microservice.

It's a good practice to keep a backup of these properties in the pet-owner resource folder as the Dockerized Consul instance may lose the configurations upon restarting.

We can review the configurations by navigating to the pet-owner application.yml file:

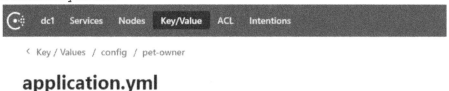

```
1  micronaut:
2    application:
3      name: pet-owner
4    router:
5      static-resources:
6        swagger:
7          paths: classpath:META-INF/swagger
8          mapping: /swagger/**
9
10 datasources:
11   default:
12     url: "jdbc:postgresql://localhost:5432/postgres"
13     username: postgres
14     password: postgres
15     driverClassName: org.postgresql.Driver
16
17 jpa:
18   default:
19     entity-scan:
20       packages:
21         - com.packtpub.micronaut.domain
22     properties:
23       hibernate:
24         show_sql: false
25         dialect: org.hibernate.dialect.PostgreSQL95Dialect
26         enable_lazy_load_no_trans: true
27
```

Figure 7.1 – Managing the pet-owner configurations in Consul

As shown in *Figure 7.1*, we can easily review/modify application configurations in Consul. We have the option to **Save** the edits or **Delete** the application.yml file (in case you want to start fresh).

Now let's dive into the changes we need to make in the pet-owner microservice to integrate it with Consul configuration management.

## Integrating the pet-owner microservice with Consul configuration management

To integrate the pet-owner microservice with Consul, we will need to add the following dependency to the pom.xml project:

```
<dependency>
    <groupId>io.micronaut</groupId>
    <artifactId>micronaut-discovery-client</artifactId>
</dependency>
```

By importing the micronaut-discover-client dependency, we can leverage out-of-the-box features for service discovery (which we will cover later in the chapter) and integration with Consul.

We will also need to add a new bootstrap.yml file in the pet-owner microservice resources directory. Bootstrap YAML will inform the service to load (or bootstrap) the application properties from an external resource while starting up. To do so, add the following configurations to the bootstrap.yml file:

```
micronaut:
  application:
    name: pet-owner
  config-client:
    enabled: true


consul:
  client:
    config:
      enabled: true
      format: yaml
      defaultZone: «${CONSUL_HOST:localhost}:${CONSUL_
PORT:8500}»
      path: "config/pet-owner/"
```

```
registration:
    enabled: true
```

All the configurations related to the Consul server are prefixed with the keyword `consul`. Some configurations to ponder are as follows:

- **defaultZone**: This parameter points to the Consul server. By default, we provide fallbacks of localhost and port `8500`.

- **path**: This points to the `application.yml` relative path on the Consul server.

These changes in `bootstap.yml` will enable the loading of the configuration for the `pet-owner` microservice from the Consul server. We can verify the changes made by running the `pet-owner` microservice. When booting up, it should sync with Consul to load the configuration:

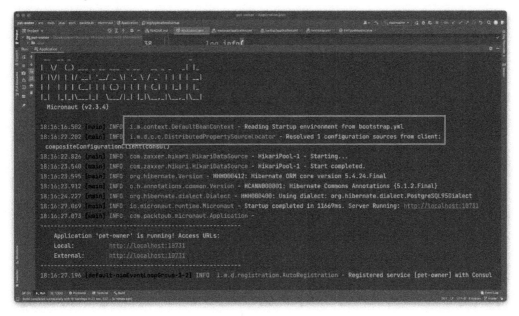

Figure 7.2 – Pet-owner microservice loading the service configurations from Consul at bootup

When we boot up the pet-owner microservice, `bootstap.yml` will inform the service to load the configuration from the Consul server. Built-in components in `micronaut-discover-client` dependency will sync with Consul and load this external configuration.

Up to here, you have learned how to employ Micronaut for distributed configuration management. In the next section, we will implement API documentation using the Micronaut framework.

# Documenting the service APIs

API documentation is especially important in a microservices architecture for intuitive access to the API information, since an application can have multiple microservices and each microservice runs several API endpoints. **OpenAPI** is the *de facto* standard for any web services documentation. The **Swagger framework**, an implementation of OpenAPI, is integrated easily with a Micronaut microservice. We can easily document all our service endpoints using Swagger. In the following sections, we will explore how we can integrate Swagger into the `pet-owner` microservice.

## Using Swagger to document the pet-owner service endpoints

To start using Swagger in the `pet-owner` microservice, we first need to import the following dependency into the `pom.xml` project:

```xml
<!-- Swagger -->
<dependency>
    <groupId>io.swagger.core.v3</groupId>
    <artifactId>swagger-annotations</artifactId>
    <version>${swagger.version}</version>
    <scope>compile</scope>
</dependency>
```

By importing the preceding dependency, we can leverage the **Swagger v3** toolbox in the pet-owner microservice. We can configure `swagger.version` in the prompt properties section. We will also need to amend the annotation processing so that the annotation processor can generate Swagger artifacts at compile time. Add the following path to the `maven-compiler-plugin` annotation processing paths:

```
<path>
  <groupId>io.micronaut.openapi</groupId>
  <artifactId>micronaut-openapi</artifactId>
  <version>2.3.1</version>
</path>
```

Adding `micronaut-openapi` to the annotation processor paths will enable `maven-compile` to build the Swagger artifacts for the project. Once you have made the changes to the project prompt, add the following annotation to the main class:

```
@OpenAPIDefinition(
    info = @Info(
        title = "pet-owner-service",
        version = "1.0",
        description = "Pet Owner APIs"
    )
)
@Singleton
public class Application {
...
}
```

Using the @OpenAPIDefinition annotation will create a Swagger artifact with the
<title>-<version>.yml pattern in the generated sources. In our case, it will create
the pet-owner-service-1.0.yml Swagger artifact when we build the project:

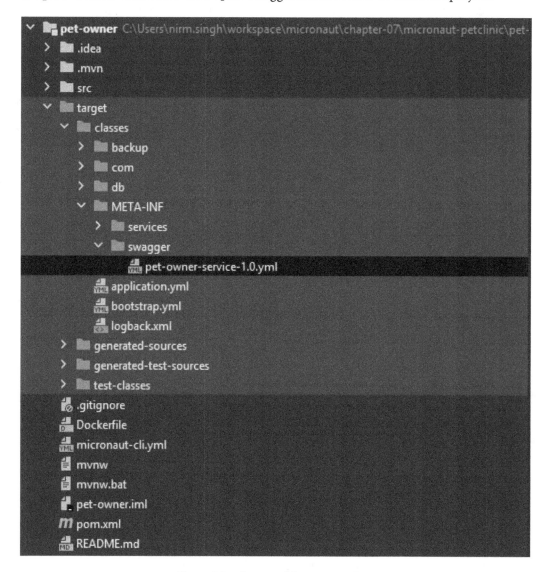

Figure 7.3 – Generated Swagger artifact

As shown in *Figure 7.3*, Swagger will create the `pet-owner-service-1.0.yml` artifact in the target folder. The generated text-only artifact can be opened in the Swagger Editor at `https://editor.swagger.io/`:

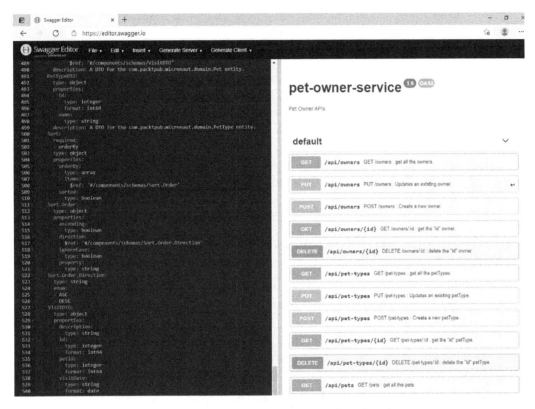

Figure 7.4 – Reviewing the Swagger YAML in the Editor

Using the Swagger Editor to review the generated YAML is very intuitive. It provides a simple user interface for all the service endpoints including the option to try out an API call. Though Micronaut provides a mechanism to generate Swagger UI views, it's very new and requires a lot of changes. Therefore, using the standard Swagger Editor is an easier and quicker option.

API documentation comes in handy when we have different product teams working on isolated microservices. Moreover, if a microservice is exposed to the end users, it's the go-to resource to know everything about a service endpoint. Continuing with this amalgamation journey, in the next section, we will implement service discovery for all the microservices in the `pet-clinic` application.

# Implementing service discovery

In the traditional monolithic architecture, if an application has multiple services then often these services are running on fixed and well-known locations (such as a URL or ports). This understanding of "well-known" is coupled into the code logic to make inter-service calls. A consumer service will call another service either at the code level or use hardcoded remote calls over the network.

By contrast, often microservices are running in virtualized or containerized environments and IP ports are assigned dynamically. To facilitate inter-service calls, we implement service discovery. In the service discovery pattern, all the microservices will register their running instances with service discovery, and clients (that is, upstream clients or even another service) will then sync up with service discovery to get the network location of the required service. Furthermore, service discovery will maintain a continuous health check on all the registered services. In the following section, we will implement service discovery using Consul in the Micronaut framework.

## Implementing service discovery using Consul

To enable service discovery, you need to import the following dependency in the pom. xml project file. We have already added this to the pet-owner microservice; now add this to the pet-clinic and pet-clinic-reviews microservices as well:

```xml
<!-- Service discovery -->
    <dependency>
        <groupId>io.micronaut</groupId>
        <artifactId>micronaut-discovery-client</artifactId>
    </dependency>
```

Once micronaut-discovery-client is imported into the services, we can leverage its service discovery capabilities.

In order to synchronize the services with Consul, make the following changes in the pet-clinic and pet-clinic-reviews microservices:

```yaml
consul:
  client:
    registration:
      enabled: true
```

By making these changes, we enable the `pet-clinic` and `pet-clinic-reviews` microservices to register with Consul service discovery. Micronaut's `micronaut-discovery-client` implementation already has the required tools so we don't need to make any code changes. To verify that all the services are registered with Consul, just run the services and they will automatically register with their application name, as shown in the following screenshot:

> **Note**
>
> If you are using **Apache Kafka** in the `pet-clinic` and `pet-clinic-reviews` microservices, then start up the Docker container for Apache Kafka before the services start up.

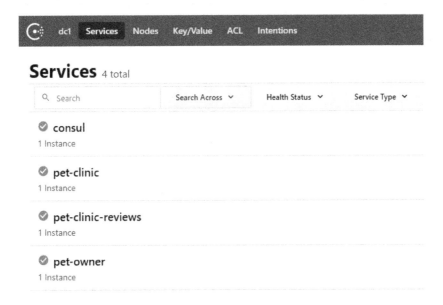

Figure 7.5 – Pet-clinic application service discovery

After successful startups, all the microservices will register their instances with the Consul service discovery. We can view the running services by going to the **Services** screen on Consul.

Though service discovery brings all the services under one umbrella by centralizing the runtime metadata (mainly the network locations), it still leaves a gap for upstream consumers as we don't have a unified interface yet. In the next section, we will implement an API gateway for the `pet-clinic` application, which will provide a unified interface for all the clients.

# Implementing the API gateway

To further the earlier discussion on dynamic network locations in the *microservices architecture*, we will focus on the API gateway now. The API gateway is an orchestrator service meant to provide unified, ubiquitous access to all the services. Though we can have multiple microservices running in the backend, an API gateway can provide a unified interface for the upstream consumers to access them. For the upstream consumers, the API gateway appears to be the only service running in the backend. On receiving a client request, the API gateway determines which service instance to call using service discovery.

In order to learn how to implement the API gateway, we will add an API gateway service to the `pet-clinic` application. Since this microservice is an orchestrator service, we can call it `pet-clinic-concierge`. The system components after the gateway service would be as seen in *Figure 7.6*:

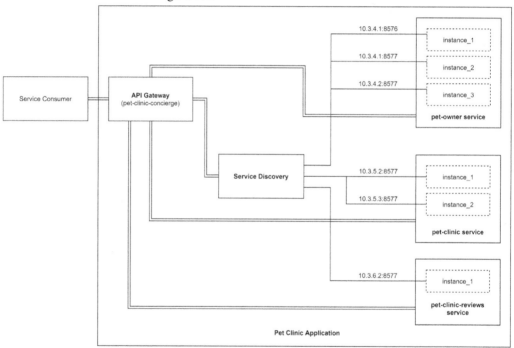

Figure 7.6 – Pet Clinic Application with service discovery and an API gateway

In the preceding diagram, we can see the **API Gateway** and **Service Discovery** orchestration in the `pet-clinic` application. The `pet-clinic-concierge` service will implement the API gateway, and any service consumer will invoke the gateway (not the service or service discovery), which will determine the service instance by synchronizing up with service discovery. The double-line connectors in the diagram show how an actual service request will be executed inside the `pet-clinic` application.

In the next section, we will dive into how to implement the API gateway in the `pet-clinic-concierge` service.

# Implementing the API gateway service

In order to learn how we can implement an API gateway in Micronaut, we will create a new service project called `pet-clinic-concierge`. To generate the boilerplate, follow the instructions:

1.  Open **Micronaut Launch** by accessing `https://micronaut.io/launch/`.
2.  Select **Micronaut Application** from the **Application Type** dropdown.
3.  Select **Java 13** from the **Java Version** dropdown.
4.  Enter `com.packtpub.micronaut` in the **Base Package** input box.
5.  Enter `pet-clinic-concierge` in the **Name** input box.
6.  Select the following features from the **Features** multi-selection options:

    **config-consul discovery-consul**

    **http-client**

    **netflix-hystrix**

    **netflix-ribbon**

    **openapi**
7.  Hit **Generate Project** button and select the **Download Zip** option.

Micronaut Launch will now generate the boilerplate for the `pet-clinic-concierge` service. In Micronaut Launch, we opted for discovery and OpenAPI, so the boilerplate will have these features already enabled and configured. In the next section, we will explore implementing a unified service façade in the `pet-clinic-concierge` service.

## Implementing a unified service façade for the API gateway

To begin, we will need to copy all the **Data Transfer Objects** (**DTOs**) to the `pet-clinic-concierge` service. These DTOs will be used in implementing clients for all the `pet-clinic` services. Copy all the DTOs in the `pet-clinic-concierge` project. We can then define the clients for all the RESTful services.

In the next section, we will focus on defining the client for the `pet-owner` microservice.

## Accessing pet-owner resources

To access the pet-owner resources, we will create clients under the com.packtpub.
micronaut.web.rest.client.petowner package. For each resource controller in
the pet-owner microservice, we will declare an HTTP client interface. The following is
the client interface for OwnerResource:

```
@Client(id = "pet-owner")
public interface OwnerResourceClient {
    @Post("/api/owners")
    HttpResponse<OwnerDTO> createOwner(@Body OwnerDTO
     ownerDTO);
    @Put("/api/owners")
    HttpResponse<OwnerDTO> updateOwner(@Body OwnerDTO
     ownerDTO);
    @Get("/api/owners")
    HttpResponse<List<OwnerDTO>> getAllOwners(HttpRequest
     request, Pageable pageable);
    @Get("/api/owners/{id}")
    Optional<OwnerDTO> getOwner(@PathVariable Long id);

    @Delete("/api/owners/{id}")
    HttpResponse deleteOwner(@PathVariable Long id);
}
```

The @Client annotation will implement a concrete client. This client will integrate with
a Consul service instance for the pet-owner service. We would need to declare all the
RESTful methods exposed in OwnerResource with their relative paths.

Although post-build we will have a concrete OwnerResourceClient, we will still need
to map the various RESTful methods in OwnerResourceClient to a local controller.
This controller will then be exposed as a service façade for the upstream consumer. For
OwnerResourceClient we can create OwnerResourceClientController
as follows:

```
@Controller("/api")
public class OwnerResourceClientController {
    @Inject
    OwnerResourceClient ownerResourceClient;
    @Post("/owners")
```

```
    public HttpResponse<OwnerDTO> createOwner(OwnerDTO
      ownerDTO) {
        return ownerResourceClient.createOwner(ownerDTO);
    }
    @Put("/owners")
    HttpResponse<OwnerDTO> updateOwner(@Body OwnerDTO
        ownerDTO) {
        return ownerResourceClient.updateOwner(ownerDTO);
    }
    @Get("/owners")
    public HttpResponse<List<OwnerDTO>>
      getAllOwners(HttpRequest request, Pageable pageable) {
        return ownerResourceClient.getAllOwners(request,
          pageable);
    }
    @Get("/owners/{id}")
    public Optional<OwnerDTO> getOwner(@PathVariable Long
      id) {
        return ownerResourceClient.getOwner(id);
    }
    @Delete("/owners/{id}")
    HttpResponse deleteOwner(@PathVariable Long id) {
        return ownerResourceClient.deleteOwner(id);
    }
}
```

In `OwnerResourceClientController`, we are injecting `OwnerResourceClient`. Any incoming request to `OwnerResourceClientController` will be passed to the client, which will then call a `pet-owner` service instance (after syncing with Consul service discovery) for further processing. Similarly, you can implement `Clients` and `Controllers` for other resources in the `pet-owner` microservice.

Next, we will implement the service façade for the `pet-clinic` resources as well.

### Accessing the pet-clinic resources

To access the `pet-clinic` resources we will create clients under the `com.packtpub.micronaut.web.rest.client.petclinic` package. For each resource controller in the `pet-clinic` microservice, we will declare an HTTP client interface. The following is the client interface for `VetResource`:

```
@Client(id = "pet-clinic")
public interface VetResourceClient {
    @Post("/api/vets")
    HttpResponse<VetDTO> createVet(@Body VetDTO vetDTO);
    @Put("/api/vets")
    HttpResponse<VetDTO> updateVet(@Body VetDTO vetDTO);
    ...
}
```

The `@Client` annotation will implement a concrete client using the `pet-clinic` service instance in service discovery. To expose these methods on a service façade, we will implement `VetResourceClientController`:

```
@Controller("/api")
public class VetResourceClientController {
    @Inject
    VetResourceClient vetResourceClient;
    @Post("/vets")
    public HttpResponse<VetDTO> createVet(VetDTO vetDTO) {
        return vetResourceClient.createVet(vetDTO);
    }
    @Put("/vets")
    public HttpResponse<VetDTO> updateVet(VetDTO vetDTO) {
        return vetResourceClient.updateVet(vetDTO);
    }
    ...
}
```

We are injecting `VetResourceClient` into `VetResourceClientController`, so any incoming request to the controller will be passed to the client, which will invoke a `pet-clinic` service instance for further processing.

In the next section, our focus will be on implementing the service façade for `pet-clinic-reviews`.

## Accessing the pet-clinic-reviews resources

For accessing `pet-clinic-reviews` resources, you will create clients under the `com.packtpub.micronaut.web.rest.client.petclinicreviews` package. We will first declare a client interface for `VetReviewResource`:

```
@Client(id = "pet-clinic-reviews")
public interface VetReviewResourceClient {
    @Post("/api/vet-reviews")
    HttpResponse<VetReviewDTO> createVetReview(@Body
      VetReviewDTO vetReviewDTO);
    @Put("/api/vet-reviews")
    HttpResponse<VetReviewDTO> updateVetReview(@Body
      VetReviewDTO vetReviewDTO);
    ...
}
```

The `@Client` annotation will implement a concrete client using the `pet-clinic-reviews` service instance in service discovery. To expose these methods on a service façade we will implement `VetReviewResourceClientController`:

```
@Controller("/api")
public class VetReviewResourceClientController {
    @Inject
    VetReviewResourceClient vetReviewResourceClient;
    @Post("/vet-reviews")
    public HttpResponse<VetReviewDTO>
      createVetReview(VetReviewDTO vetReviewDTO) {
        return vetReviewResourceClient.createVetReview
          (vetReviewDTO);
    }
    @Put("/vet-reviews")
    public HttpResponse<VetReviewDTO>
      updateVetReview(VetReviewDTO vetReviewDTO) {
        return vetReviewResourceClient.updateVetReview
          (vetReviewDTO);
    }
    ...
}
```

Here we inject `VetReviewResourceClient` into `VetReviewResourceClientController` and incoming requests to the controller will be passed to the client, which we will invoke on a `pet-clinic-reviews` service instance for further processing.

In the next section, we will focus on handling fault tolerance concerns in regard to the microservices.

# Implementing the fault tolerance mechanisms

Faults and failures are inevitable in a microservice environment. As the numbers of distributed components increase, the number of faults both within each component and those originating from their interactions increase as well. Any microservices application must have built-in resilience for these unfortunate scenarios. In this section, we will explore and implement different ways to handle faults and failures in the Micronaut framework.

## Leveraging built-in mechanisms

Micronaut is a cloud-native framework and has built-in capabilities to handle faults and failures. Essentially, its fault tolerance is driven by **retry advice**. Retry advice offers the `@Retryable` and `@CircuitBreaker` annotations that can be used in any HTTP client.

### Using @Retryable in an HTTP client

**@Retryable** is an effortless but effective fault tolerance mechanism – to put it simply, it's used to try again in case of a failure. These try attempts can be made again after a fixed delay and can continue until the service either responds back or no more attempts are left.

To use `@Retryable` we can just annotate the client declaration. We can use `@Retryable` on `OwnerResource` as follows:

```
@Retryable(attempts = "5", delay = "2s", multiplier = "1.5",
maxDelay = "20s")
@Client(id = "pet-owner")
public interface OwnerResourceClient {

   ...

}
```

By using `@Retryable` on `OwnerResourceClient` you enable fault-tolerance on all its methods. If the `pet-owner` microservice is down then `OwnerResourceClient` will retry attempting to establish the communication for a maximum of five attempts. We can configure `@Retryable` with the following settings:

- **attempts**: This denotes the maximum number of retries the client can make. By default, this is 3 attempts.
- **delay**: This marks the delay between the retries and is 1 second by default.
- **multiplier**: This specifies the multiplier used to calculate the delay and is 1.0 by default.
- **maxDelay**: This specifies the maximum overall delay and is empty by default. If specified, any retry attempts will be halted if retries reach the maximum delay limit.

`@Retryable` is well suited for temporary/momentary faults but for long-lasting faults, we would need to use the circuit breaker pattern. In the next section, we will see how we can use the circuit breaker in the Micronaut framework.

## Using @CircuitBreaker in an HTTP client

As we discussed, failure in a highly distributed system such as microservices is inevitable. In a microservice architecture, if a service goes down there is another defensive mechanism to help avoid congesting the service traffic with more requests until the service is healthy again. This mechanism is called the **circuit breaker**. In normal circumstances, the circuit is open and accepting requests. On a failure instance, a counter is increased until it reaches a specified threshold. After reaching the threshold, the circuit is put into a closed state and the service will respond immediately with an error avoiding any timeouts. The circuit breaker has an internal polling mechanism to determine the health of the service and if the service is healthy again then the circuit is put back to an open state.

We can simply use Micronaut's built-in annotation to specify a circuit breaker on an HTTP client. Let's implement a circuit breaker in `PetResourceClient`:

```
@Client(id = "pet-owner")
@CircuitBreaker(delay = "5s", attempts = "3", multiplier = "2",
reset = "300s")
public interface PetResourceClient {
    @Post("/api/pets")
    HttpResponse<PetDTO> createPet(@Body PetDTO petDTO);
```

```
@Put("/api/pets")
HttpResponse<PetDTO> updatePet(@Body PetDTO petDTO);
...
}
```

In the preceding circuit breaker implementation, if the `PetResource` endpoints fail then `PetResourceClient` will try five attempts, waiting for 3 seconds for the first attempt and a multiplier of two for future attempts. After five attempts, if the service is still not responding, then the circuit will be put into a closed state. It will try to access it again after a reset interval of 5 minutes to check if the service is healthy yet.

## Using @Fallback for an HTTP client

Often in a circuit breaker implementation, there are **feign clients** or **fallbacks**. Instead of raising a server error when the circuit is closed, a fallback implementation can handle the request and respond normally. This is especially effective when the actual service call might be returning that a fallback can also return.

In the following example of the circuit breaker in `PetResourceClient`, we can create a simple fallback that will handle the incoming requests when the circuit is closed:

```
@Fallback
public class PetResourceFallback implements PetResourceClient {
    @Override
    public HttpResponse<PetDTO> createPet(PetDTO petDTO) {
        return HttpResponse.ok();
    }
    @Override
    public HttpResponse<PetDTO> updatePet(PetDTO petDTO) {
        return HttpResponse.ok();
    }
    @Override
    public HttpResponse<List<PetDTO>>
      getAllPets(HttpRequest request, Pageable pageable) {
        return HttpResponse.ok();
    }
    @Override
    public Optional<PetDTO> getPet(Long id) {
        return Optional.empty();
    }
```

```
    @Override
    public HttpResponse deletePet(Long id) {
        return HttpResponse.noContent();
    }
}
```

PetResourceFallback has default implementations of all the PetResource endpoints and it will provide a gracious response when PetResource is unaccessible. In this example, we return an empty response from all the endpoints. You can tinker with the implementation and create a default response as desired.

# Summary

In this chapter, you learned how to handle various microservice concerns in the Micronaut framework. We kickstarted the journey by externalizing the application configurations using Consul and learned why distributed configuration management is required in the microservices. We then dived into how to automate API documentation using OpenAPI and Swagger. Later, we discussed service discovery and the API gateway and implemented those in the pet clinic application.

At last, we explored the need for fault tolerance and how you can simply use the built-in mechanism in the Micronaut framework for building resilience in your microservices application.

This chapter has equipped you with all the first-hand knowledge you require to handle various microservice concerns relating to service discovery, API gateways, and fault tolerance. The chapter took a practical approach by adding to the pet-clinic application while covering these aspects.

In the next chapter, you will explore various ways to deploy the pet-clinic microservice application.

# Questions

1. What is distributed configuration management?

2. How can you implement a configuration store in Consul?

3. How can you automate the process of API documentation using Swagger in the Micronaut framework?

4. What is service discovery?

5. How can you implement service discovery in the Micronaut framework?

6. What is an API gateway in the microservices architecture?

7. How can you implement an API gateway in the Micronaut framework?

8. What is `@Retryable` in Micronaut?

9. What is `@CircuitBreaker` in Micronaut?

10. How can you implement a circuit breaker in the Micronaut framework?

11. How can you implement a fallback in Micronaut?

# 8
# Deploying Microservices

The literal meaning of *deployment* is to bring resources into effective action. Therefore, in the microservices context, it means to bring microservices into effective action. Any service deployment is a multi-step process and often involves building the artifacts and then pushing the artifacts to a runtime environment. In the microservices world, an effective strategy for microservices deployment is crucial. Essentially, we need to watch out for the following when planning a deployment process:

- Continue with the pattern of *separation of concern* and self-isolate the artifact-building process for each microservice.

- Decouple any connection requirements within microservices and let the service discovery or an implementation close to service discovery handle the microservice bindings.

- Implement a seamless deployment process that can handle instantiating all of the microservice application components in a unified and automated way.

In this chapter, we will dive into these aforementioned concerns while covering the following topics:

- Building the container artifacts
- Deploying the container artifacts

By the end of this chapter, you will be well versed in these aspects of microservices deployment.

# Technical requirements

All the commands and technical instructions in this chapter run on Windows 10 and macOS. The code examples covered in this chapter are available in the book's GitHub repo here:

```
https://github.com/PacktPublishing/Building-Microservices-
with-Micronaut/tree/master/Chapter08
```

The following tools need to be installed and set up in the development environment:

- **Java SDK**: Version 13 or above (we used Java 14).

- **Maven**: This is optional and only required if you would like to use Maven as the build system. However, we recommend having Maven set up on any development machine. Instructions to download and install Maven can be found at `https://maven.apache.org/download.cgi`.

- **A development IDE**: Based on your preference, any Java-based IDE can be used, but for the purpose of writing this chapter, IntelliJ was used.

- **Git**: Instructions to download and install this can be found at `https://git-scm.com/downloads`.

- **PostgreSQL**: Instructions to download and install this can be found at `https://www.postgresql.org/download/`.

- **MongoDB**: MongoDB Atlas provides a free online database-as-a-service with up to 512 MB storage. However, if a local database is preferred then instructions to download and install it can be found at `https://docs.mongodb.com/manual/administration/install-community/`. We used a local installation for writing this chapter.

- **A REST client**: Any HTTP REST client can be used. We used the **Advanced REST Client** Chrome plugin.

- **Docker**: Instructions to download and install Docker can be found at `https://docs.docker.com/get-docker/`.

# Building the container artifacts

To deploy any application, the first step is to build all the required artifacts. And building an artifact typically involves checking out the source code and compiling and creating a deployable artifact. In regard to microservices, often, this deployable form is a Docker container image. Docker images have elegantly decoupled the runtime topology requirements. A Docker image is platform agnostic and can be deployed to any host machine that runs Docker. In the next section, we will dive into how to build Docker images for the microservices in our pet-clinic application.

## Containerizing the Micronaut microservices using Jib

Jib is a containerizing framework from Google that can seamlessly tie with Java build frameworks such as Maven or Gradle to build container images. Jib has hugely simplified the process to create container (Docker) images. Let's quickly see the workflow without Jib to create a Docker image:

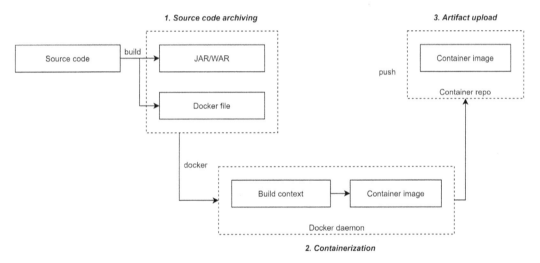

Figure 8.1 – Docker containerization workflow without Jib

As you can see in the preceding diagram, to build a Docker image without Jib we need to build the **Source code** and create a **Docker file**. The **Docker daemon** then uses the build context to create a **Container image** and push it to the repository/registry.

In contrast to the preceding workflow, Jib simplifies the whole process as shown in the following diagram:

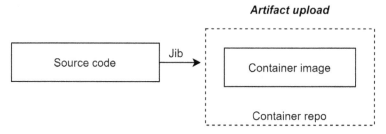

Figure 8.2 – Docker containerization with Jib

Jib takes the configuration from the build file such as project's pom file or a Gradle file to build the container image and uploads it to the configured repository/registry.

We will use Jib for containerizing our microservices in the following sections.

## Containerizing the pet-owner microservice

In the pet-owner microservice, we will simplify configuration management to use local `application.properties`. We can still use `consul` for configuration management, but to focus on the containerization aspect, we opt to use a local `application.properties` file. Delete or back up the `bootstrap.yml` file and make the following changes to the `application.properties` file:

```
micronaut:
  application:
    name: pet-owner
  server:
    port: 32581
  router:
    static-resources:
      swagger:
        paths: classpath:META-INF/swagger
        mapping: /swagger/**

datasources:
  default:
    url: "jdbc:postgresql://host.docker.internal:5432/postgres"
    username: postgres
    password: postgres
```

```
        driverClassName: org.postgresql.Driver

consul:
  client:
    default-zone: "consul:8500"
    registration:
      enabled: true

jpa:
  ...
```

A few things to ponder in these application properties changes are as follows:

- **Datasource URL**: We will be using `host.docker.internal` instead of localhost to point to the Postgres instance installed on the host operating system (outside Docker).

- **Consul default-zone**: In our Docker services, we will configure a `consul` service. To use Dockerized `consul` from a pet-owner Docker container, we will need to specify the service name instead of localhost.

- **Port**: We are specifying a fixed port to run the pet-owner microservice as this will expose this port later in the deployment.

Essentially, to use anything from the host machine we should use `host.docker.internal`, and to use any Docker service container we must specify the Docker service name.

After making the previous application configuration changes, we can proceed to containerization. To containerize the pet-owner microservice we will use Jib. Make the following changes in the project's pom file build settings:

```
<build>
  <plugins>
    ...
    <plugin>
      <groupId>com.google.cloud.tools</groupId>
      <artifactId>jib-maven-plugin</artifactId>
      <version>2.8.0</version>
      <configuration>
        <from>
```

```
        <image>openjdk:13-jdk-slim</image>
      </from>
      <to>
        <image>pet-owner-0.1-image</image>
      </to>
      <container>
        <creationTime>${maven.build.timestamp}</creationTime>
      </container>
    </configuration>
  </plugin>
 </plugins>
</build>
```

In the preceding pom changes, we are using `jib-maven-plugin` to build the container image. The `<configuration>` section specifies the Docker configurations such as the `<from>` image (which is pointing to use JDK 13). To name the created image, we use `<to>` along with `<creationTime>` to correctly stamp the time coordinates on the image.

To build the image, perform the following steps:

1. Open the terminal and change directory to the `pet-owner` root directory.
2. Type and run the `mvn compile jib:dockerBuild` command in the terminal.
3. Wait for the command to finish.

These instructions will create a local Docker image that can be verified using the `docker images | grep pet-owner` command in the terminal:

```
nirm.singh@CA-L2WT60G2 MINGW64 ~/workspace/micronaut/Chapter-08/micronaut-petclinic
$ docker images | grep pet-owner
pet-owner-0.1-image          latest          a65b932fde5a    24 hours ago    441MB
```

Figure 8.3 – Verifying the pet-owner image in the local Docker registry

In the preceding screenshot, we can see the output of `docker images`. A `pet-owner-0.1-image` image is stored in the local Docker registry.

## Containerizing the pet-clinic microservice

To make the pet-clinic microservice container ready, make the following changes to the `application.properties` file:

```
micronaut:
  application:
```

```
      name: pet-clinic
   server:
     port: 32582

kafka:
  bootstrap:
     servers: kafka:9092

datasources:
  default:
    url:
      «jdbc:postgresql://host.docker.internal:5432/postgres»
    username: postgres
    password: postgres
    driverClassName: org.postgresql.Driver

consul:
  client:
    default-zone: "consul:8500"
    registration:
       enabled: true
  …
```

The following are a few things to ponder in application properties changes:

- **Datasource URL**: We will be using `host.docker.internal` instead of localhost to point to Postgres installed in the host operating system (outside Docker).

- **Consul default-zone**: In our Docker services, we will configure a `consul` service. To use Dockerized `consul` from the pet-clinic Docker container, we will need to specify the service name instead of localhost.

- **Kafka server**: In our Docker services, we will configure a **Kafka** service.

- **Port**: We are specifying a fixed port to run the pet-clinic microservice as we will expose this port later in the deployment.

After making the preceding application configuration changes, we can proceed to containerization. To containerize the pet-clinic microservice we will use Jib. Make the following changes in the project's pom file build settings:

```
<build>
  <plugins>
    ...
    <plugin>
      <groupId>com.google.cloud.tools</groupId>
      <artifactId>jib-maven-plugin</artifactId>
      <version>2.8.0</version>
      <configuration>
        <from>
          <image>openjdk:13-jdk-slim</image>
        </from>
        <to>
          <image>pet-clinic-0.1-image</image>
        </to>
        <container>
          <creationTime>${maven.build.timestamp}</creationTime>
        </container>
      </configuration>
    </plugin>
  </plugins>
</build>
```

In the preceding pom changes, we use jib-maven-plugin to build the container image. The <configuration> section specifies Docker configurations such as the <from> image (which is pointing to use JDK 13). To name the created image we use <to> along with <creationTime> to correctly stamp the time coordinates on the image.

To build the image, we perform the following steps:

1. Open the terminal and change directory to the pet-clinic root directory.

2. Type and run the mvn compile jib:dockerBuild command in the terminal.

3. Wait for the command to finish.

The preceding instructions will create a local Docker image that can be verified using the docker images | grep pet-clinic command in the terminal.

## Containerizing the pet-clinic-reviews microservice

To make the pet-clinic-reviews microservice container ready, make the following changes to the `application.properties` file:

```
micronaut:
  application:
    name: pet-clinic-reviews
  server:
    port: 32583

kafka:
  bootstrap:
    servers: kafka:9092

mongodb:
  uri: mongodb://mongodb:mongodb@host.docker.internal:27017/
pet-clinic-reviews
    databaseName: pet-clinic-reviews
    collectionName: vet-reviews

consul:
  client:
    default-zone: «consul:8500»
    registration:
      enabled: true
...
```

The following are a few things to ponder in application properties changes:

- **Mongo database URI**: We will be using `host.docker.internal` instead of localhost to point to the Mongo DB instance installed on the host operating system (outside Docker).

- **Consul default-zone**: In our Docker services, we will configure a `consul` service. To use Dockerized `consul` from the `pet-clinic-reviews` Docker container, we will need to specify the service name instead of localhost.

- **Kafka server**: In our Docker services, we will configure a `kafka` service. To use Dockerized `consul` from the `pet-clinic-reviews` Docker container, we will need to specify the service name instead of localhost.

- **Port**: We are specifying a fixed port to run the pet-clinic-reviews microservice as we will expose this port later in the deployment.

After making the preceding application configuration changes, we can proceed to containerization. To containerize the `pet-clinic-reviews` microservice we will use Jib. Make the following changes in the project's pom fie build settings:

```xml
<build>
  <plugins>
    ...
    <plugin>
      <groupId>com.google.cloud.tools</groupId>
      <artifactId>jib-maven-plugin</artifactId>
      <version>2.8.0</version>
      <configuration>
        <from>
          <image>openjdk:13-jdk-slim</image>
        </from>
        <to>
          <image>pet-clinic-reviews-0.1-image</image>
        </to>
        <container>
          <creationTime>${maven.build.timestamp}</creationTime>
        </container>
      </configuration>
    </plugin>
  </plugins>
</build>
```

In the preceding pom changes, we are using `jib-maven-plugin` to build the container image. The `<configuration>` section specifies Docker configurations such as the `<from>` image (which is pointing to use JDK 13). To name the created image, we use `<to>` along with `<creationTime>` to correctly stamp the time coordinates on the image.

To build the image, perform the following steps:

1.  Open the terminal and change directory to the `pet-clinic-reviews` root directory.

2.  Type and run the `mvn compile jib:dockerBuild` command in terminal.

3.  Wait for the command to finish.

The preceding instructions will create a local Docker image that can be verified using the `docker images | grep pet-clinic-reviews` command in the terminal.

## Containerizing the pet-clinic-concierge microservice

To make the pet-clinic-concierge microservice container ready, make the following changes to the `application.properties` file:

```
micronaut:
  application:
    name: pet-clinic-concierge
  server:
    port: 32584
  config-client:
    enabled: true


consul:
  client:
    default-zone: "consul:8500"
    registration:
      enabled: true
...
```

The following are a few things to ponder in application properties changes:

*   **Consul default-zone**: In our Docker services, we will configure a `consul` service. To use Dockerized `consul` from the pet-clinic-concierge Docker container we will need to specify the service name instead of localhost.

*   **Port**: We are specifying a fixed port to run the `pet-clinic-concierge` microservice as we will expose this port later in the deployment.

After making the preceding application configuration changes we can proceed to containerization. To containerize the pet-clinic-concierge microservice (the API gateway) we will use jib. Make the following changes in the project's pom file build settings:

```
<build>
  <plugins>
    ...
    <plugin>
      <groupId>com.google.cloud.tools</groupId>
      <artifactId>jib-maven-plugin</artifactId>
      <version>2.8.0</version>
      <configuration>
        <from>
          <image>openjdk:13-jdk-slim</image>
        </from>
        <to>
          <image>pet-clinic-concierge-0.1-image</image>
        </to>
        <container>
          <creationTime>${maven.build.timestamp}</creationTime>
        </container>
      </configuration>
    </plugin>
  </plugins>
</build>
```

In the preceding pom changes, we use jib-maven-plugin to build the container image. The <configuration> section specifies Docker configurations such as the <from> image (which is pointing to use JDK 13). To name the created image we use <to> along with <creationTime> to correctly stamp the time coordinates on the image.

To build the image, perform the following steps:

1.  Open the terminal and change directory to the `pet-clinic-concierge` root directory.

2.  Type and run the `mvn compile jib:dockerBuild` command in terminal.

3.  Wait for the command to finish.

The preceding instructions will create a local Docker image that can be verified using the `docker images | grep pet-clinic-concierge` command in the terminal.

Now, we have containerized all the microservices. We can verify these images in the Docker Dashboard as well:

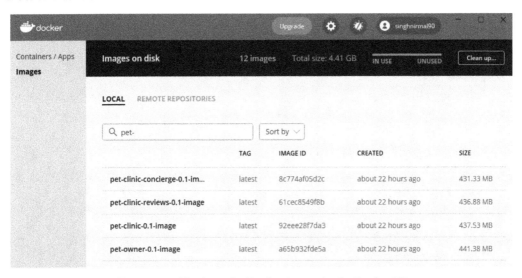

Figure 8.4 – Verifying the Docker images in the Docker UI

In the Docker UI, we can simply go to **Images** and under the **LOCAL** tab, we can filter all the pet-clinic application images. In the next section, we will use these Docker images in our pet-clinic application deployment.

# Deploying the container artifacts

In the previous section, we explored how we can use `Jib` to simplify microservice containerization. In this section, we will dive into how we can make the end-to-end deployment seamless and unified using Docker orchestration with `docker-compose`.

# Using docker-compose to deploy the pet-clinic services

docker-compose is a tool available under the Docker ecosystem and is very intuitive in defining and deploying a multi-container application. With a simple YAML-flavored syntax, we can set up all the services and their dependencies and use a single command to deploy the whole application. We will create a docker-compose file for the pet-clinic application covering all the necessary services/components, including microservices, service discovery, and the Apache Kafka ecosystem.

Firstly, let's define the ancillary services in docker-compose as follows:

```
version: '3'

services:
  consul:
    image: bitnami/consul:latest
    ports:
      - '8500:8500'

  zookeeper:
    image: bitnami/zookeeper:3-debian-10
    ports:
      - 2181:2181
    volumes:
      - zookeeper_data:/pet-clinic-reviews
    environment:
      - ALLOW_ANONYMOUS_LOGIN=yes

  kafka:
    image: bitnami/kafka:2-debian-10
    ports:
      - 9094:9094
    volumes:
      - kafka_data:/pet-clinic-reviews
    environment:
      - KAFKA_BROKER_ID=1
      - KAFKA_CFG_ZOOKEEPER_CONNECT=zookeeper:2181
      - ALLOW_PLAINTEXT_LISTENER=yes
      - KAFKA_LISTENERS=INTERNAL://kafka:9092,OUTSIDE://
```

```
kafka:9094
      - KAFKA_ADVERTISED_LISTENERS=INTERNAL://
kafka:9092,OUTSIDE://localhost:9094
      - KAFKA_LISTENER_SECURITY_PROTOCOL_
MAP=INTERNAL:PLAINTEXT,OUTSIDE:PLAINTEXT
      - KAFKA_INTER_BROKER_LISTENER_NAME=INTERNAL
    depends_on:
      - zookeeper

  kafdrop:
    image: obsidiandynamics/kafdrop
    ports:
      - 9100:9000
    environment:
      - KAFKA_BROKERCONNECT=kafka:9092
      - JVM_OPTS=-Xms32M -Xmx64M
    depends_on:
      - kafka
  ...
```

From the previous code, we see the docker-compose file. We begin by defining
a service for the consul service discovery. We will expose consul on port 8500.
Furthermore, we will define services for the Apache Kafka ecosystem; that is, Zookeeper,
Kafka, and the Kafdrop UI. Once these services are defined in the docker-compose file,
we can proceed to the pet-clinic microservices. Refer to the following code:

```
...
pet-owner:
    image: "pet-owner-0.1-image"
    ports:
        - "32581:32581"
    depends_on:
        - consul

  pet-clinic:
    image: "pet-clinic-0.1-image"
    ports:
        - "32582:32582"
```

```
        depends_on:
            - kafka
            - consul

    pet-clinic-reviews:
        image: "pet-clinic-reviews-0.1-image"
        ports:
            - "32583:32583"
        depends_on:
            - kafka
            - consul

    pet-clinic-concierge:
        image: "pet-clinic-concierge-0.1-image"
        ports:
            - "32584:32584"
        depends_on:
            - consul
...
```

While defining the configurations for the pet-clinic microservices, we can specify the dependencies using depends_on. This will ensure that Docker instantiates the services as per the dependency order. Also, for deploying each service, we will be using the pet-clinic microservices Docker images.

Once the docker-compose file is defined for the pet-clinic application, refer to the following instructions to deploy the pet-clinic application:

1. Open the bash terminal.

2. Change the directory to the location where the docker-compose file is stored.

3. Type and run the docker compose up command.

4. Wait for Docker to instantiate the containers as specified in the docker-compose file.

After the successful run of the docker-compose command, we can verify the pet-clinic application in the Docker Dashboard, shown as follows:

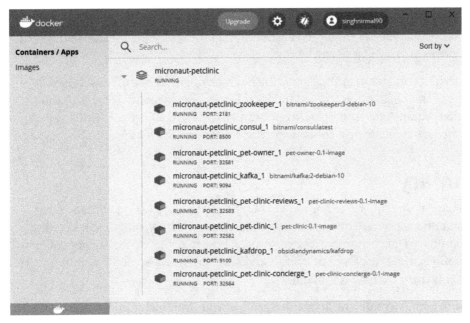

Figure 8.5 – Verifying the deployment of pet-clinic on the Docker Dashboard

In the preceding screenshot, you can see the status of all the services in the pet-clinic application. You can click on a service and monitor the logs and access the web interface (if any). Furthermore, we can check the consul service discovery for the health of the pet-clinic microservices. Refer to the following screenshot:

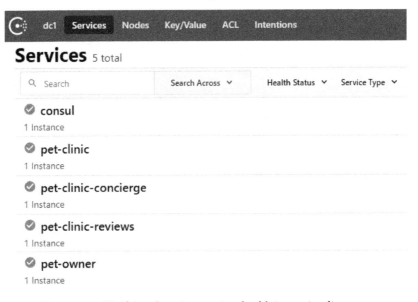

Figure 8.6 – Verifying the microservices health in service discovery

In the `consul` service discovery, we can observe the health of each instance of a microservice. In the preceding screenshot, we can see all the `pet-clinic` microservices are running normally.

Microservice container orchestration is the bedrock of any deployment strategy. For the scope of this chapter, we covered the local Docker deployment of the pet-clinic application, but the container images built can be deployed anywhere, be it locally or in a cloud environment, using a container orchestration tool such as `docker-compose`.

# Summary

In this chapter, we jumpstarted our understanding with a discussion about making Micronaut microservices container ready. Later, we dived into using Jib for creating the container images for each microservice. We saw how to define all the service container configurations using `docker-compose` and seamlessly deploy all the required service components using a single command.

This chapter enhances the deployment aspects of your Micronaut microservices journey by equipping you with first-hand knowledge on containerization and automated deployment. This skill set is much sought after in microservices application development and maintenance.

In the next chapter, we will explore various ways to monitor different aspects of the pet-clinic application in Micronaut.

# Questions

1. What is Jib?

2. How can we use Jib to create a Docker container in Micronaut?

3. How can we connect to localhost from a Docker container in Micronaut?

4. How can we deploy a multi-service application using `docker-compose` in Micronaut?

5. How can we perform Docker containerization of a Micronaut microservices application?

# Section 5: Microservices Maintenance

This section will cover the maintenance aspects of microservices in the Micronaut framework and has the following chapter:

- *Chapter 9, Distributed Logging, Tracing, and Monitoring*

# 9
# Distributed Logging, Tracing, and Monitoring

A microservice application often runs multiple microservices on a varied range of multiple hosts. For upstream consumers, the API gateway provides a one-stop-shop interface to access all the application endpoints. Any request to the API gateway is dispersed to one or more microservices. This distributed diffusion of request handling escalates challenges in maintaining microservices-based applications. If any anomaly or error occurs, it is hard to dig which microservice or distributed component is at fault. In addition, any effective microservices implementation must handle the maintenance challenges proactively.

In this chapter, we will explore the following topics:

- **Distributed logging**: How we can implement the log aggregation for distributed microservices so that application logs can be accessed and indexed in one place?

- **Distributed tracing**: How we can trace the execution of a user request that could be dispersed onto multiple microservices running on multiple host environments?

- **Distributed monitoring**: How we can continuously monitor the key performance indicators for all the service components to get a holistic picture of the system's health?

By collecting these three different kinds of data – logging, tracing, and monitoring – we enhance the system observability. By accessing this telemetry data at any point in time, we can intuitively and precisely get a complete context of what and how a request was executed in the system. To learn more about observability, we will explore the microservices patterns on distributed logging, tracing, and monitoring with the hands-on `pet-clinic` application.

By the end of this chapter, you will have good knowledge of implementing these observability patterns in the Micronaut framework.

# Technical requirements

All the commands and technical instructions in this chapter are run on Windows 10 and macOS. Code examples covered in this chapter are available in the book's GitHub repository at `https://github.com/PacktPublishing/Building-Microservices-with-Micronaut/tree/master/Chapter09`.

The following tools need to be installed and set up in the development environment:

- **Java SDK**: Version 13 or above (we used Java 14).

- **Maven**: This is optional and only required if you would like to use Maven as the build system. However, we recommend having Maven set up on any development machine. Instructions to download and install Maven can be found at `https://maven.apache.org/download.cgi`.

- **Development IDE**: Based on your preferences, any Java-based IDE can be used, but for the purpose of writing this chapter, IntelliJ was used.

- **Git**: Instructions to download and install Git can be found at `https://git-scm.com/downloads`.

- **PostgreSQL**: Instructions to download and install PostgreSQL can be found at `https://www.postgresql.org/download/`.

- **MongoDB**: MongoDB Atlas provides a free online database-as-a-service with up to 512 MB storage. However, if a local database is preferred, then instructions to download and install can be found at `https://docs.mongodb.com/manual/administration/install-community/`. We used a local installation for writing this chapter.

- **REST client**: Any HTTP REST client can be used. We used the Advanced REST Client Chrome plugin.

- **Docker**: Instructions to download and install Docker can be found at `https://docs.docker.com/get-docker/`.

# Distributed logging in Micronaut microservices

As we discussed in the chapter introduction, in a microservices-based application, a user request is executed on multiple microservices running on different host environments. Therefore, the log messages are spread across multiple host machines. This brings a unique challenge to a developer or admin maintaining the application. If there's a failure, then it will be hard to zero down on the issue as you have to sign into multiple host machines/environments, grep the logs, and put them together to make sense.

In this section, we will dive into log aggregation for distributed logging in microservices.

Log aggregation, as the name suggests, is combining the logs produced by various microservices and components in the application. Log aggregation typically involves the following components:

- **Log producer**: This is any microservice or a distributed component that's producing logs while executing the control flow.

- **Log dispatcher**: The log dispatcher is responsible for collecting the logs produced by the log producer and dispatching them to the centralized storage.

- **Log storage**: The log storage persists and indexes the logs produced by all the application components and microservices.

- **Log visualizer**: The log visualizer provides a user interface for accessing, searching, and filtering the logs stored in the log storage.

In the pet-clinic application context, we will implement the **ELK** (short for **Elasticsearch**, **Logstash**, **Kibana**) Stack for distributed logging. Refer to the following figure:

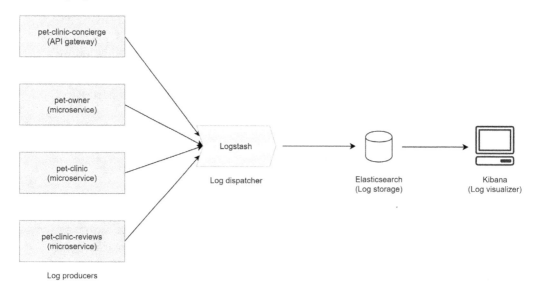

Figure 9.1 – Distributed logging using the ELK Stack

In the preceding diagram, the ELK Stack is used to implement distributed logging in the pet-clinic application. **Logstash** dispatches the logs into the **Elasticsearch** engine, which is then used by **Kibana** to provide a user interface.

In the next section, we will explore how we can set up an ELK Stack in a dockerized container.

## Setting up ELK in Docker

To set up ELK in Docker, follow these instructions:

1. Check out docker-elk from https://github.com/PacktPublishing/ Building-Microservices-with-Micronaut/tree/master/ Chapter09/micronaut-petclinic/docker-elk.

2. Open any Bash terminal (we used Git Bash).

3. Change directory to where you have checked out docker-elk.

4.  Run the `docker compose up -d` command.

5.  Wait for Docker to download the images and instantiate the ELK container.

The preceding instructions will boot up an ELK app in Docker. You can verify the installation by going to the Docker Dashboard | **Containers / Apps**, as shown here:

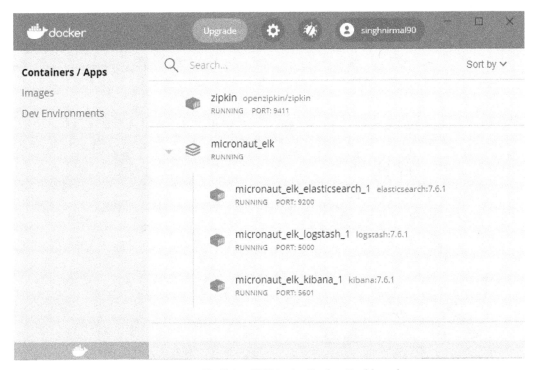

Figure 9.2 – Verifying ELK in the Docker Dashboard

Here, you can verify ELK instantiation. By default, Elasticsearch runs on port `9200`, Logstash on `5000`, and Kibana on port `5601`.

In the next section, we will modify our `pet-clinic` microservice to dispatch the logs to the Logstash instance.

## Integrating Logstash with Micronaut microservices

To integrate Logstash into the `pet-clinic` microservice, we will leverage Logback. We will introduce a new appender to Logback that can dispatch the logs to the previously created Logstash instance.

In the locally checked out `docker-elk` directory, you can verify that Logstash is configured with the following settings:

```
input {
    tcp {
        port => 5000
        type => syslog
        codec => json_lines
    }
}
filter {
    grok {
        match => [ "message", "%{GREEDYDATA}" ]
    }
    mutate {
        add_field => { "instance_name" => "%{app_name}-
          %{host}:%{app_port}" }
    }
}
output {
    stdout { # This will log all messages so that we can
      confirm that Logstash is receiving them
        codec => rubydebug
    }
    elasticsearch {
        hosts => [
          "${XPACK_MONITORING_ELASTICSEARCH_HOSTS}" ]
        user =>
          "${XPACK_MONITORING_ELASTICSEARCH_USERNAME}"
        password =>
          "${XPACK_MONITORING_ELASTICSEARCH_PASSWORD}"
        index => "logstash-%{+YYYY.MM.dd}"
    }
}
```

In `logstash.config`, we have the following three sections:

- `input`: Logstash has the power to aggregate more than 50 different kinds of log sources. `input` configures Logstash for one or more input sources. In our configuration, we are enabling `tcp` input on port `5000`.

- `filter`: Logstash's `filter` provides an easy way to transform the incoming logs into filter-defined log events. These events are then pushed to log storage. In the preceding configuration, we are using a `grok` filter along with `mutate` to add extra information (`app_name` and `app_port`) to the log events.

- `output`: The `output` section configures the receiving sources so that Logstash can push the log events to the configured output sources. In the preceding configuration, we are configuring the standard output and Elasticsearch to receive the produced log events.

So far, we have booted up an ELK Docker instance with Logstash configured to receive, transform, and send log events into Elasticsearch. Next, we will make required amends in the `pet-clinic` microservices so that logs can be shipped to Logstash.

## Configuring microservices for distributed logging

In order to make the `pet-clinic` microservice aggregate and ship logs to Logstash, we need to add the following `logstash-logback-encoder` dependency to all the microservice `pom.xml` files:

```
<dependency>
    <groupId>net.logstash.logback</groupId>
    <artifactId>logstash-logback-encoder</artifactId>
    <version>6.3</version>
</dependency>
```

By importing `logstash-logback-encoder`, we can leverage the `net.logstash.logback.appender.LogstashTcpSocketAppender` class in `logback.xml`. This class provides the `logstash` appender, which can ship logs to the Logstash server from the microservice.

Modify `logback.xml` for all microservices by adding the Logstash appender as follows:

```xml
<appender name="logstash" class="net.logstash.logback.appender.
LogstashTcpSocketAppender">
    <param name="Encoding" value="UTF-8"/>
    <remoteHost>host.docker.internal</remoteHost>
    <port>5000</port>
    <encoder
      class=»net.logstash.logback.encoder.LogstashEncoder»/>
</appender>
...
<root level="debug">
    <appender-ref ref="logstash"/>
    <appender-ref ref="stdout"/>
</root>
```

The Logstash appender will help in shipping the logs to `localhost:5000` and as we are running Logstash in a Docker container, we provide the address as `host.docker.internal`.

Also, we need to add the appender to the root level by using `appender-ref`.

Furthermore, we need to define two properties for `app_name` and `app_port`. These are the filter configurations that will be used by Logstash to create the desired log events with app information. This is how we do it:

```xml
<property scope="context" name="app_name" value="pet-owner "/>
<property scope="context" name="app_port" value="32581"/>
```

In the preceding code snippet, we have added the required properties for the `pet-owner` microservice. We need to add similar properties in all the services so Logstash can generate service-specific log events.

# Verifying the distributed logging in the pet-clinic application

To verify that Logstash is receiving the logs from all the microservices in the pet-clinic application, we would need to re-build the Docker images and redeploy the pet-clinic application. Perform the following steps:

1.  Open the terminal in the pet-owner microservice root directory:

    a. Run the jib command to build Docker mvn compile jib:dockerBuild.

    b. Wait for jib to build and upload the Docker image to the local Docker images repository.

2.  Open the terminal in the pet-clinic microservice root directory:

    a. Run the jib command to build Docker mvn compile jib:dockerBuild.

    b. Wait for jib to build and upload the Docker image to the local Docker images repository.

3.  Open the terminal in the pet-clinic-reviews microservice root directory:

    a. Run the jib command to build Docker mvn compile jib:dockerBuild.

    b. Wait for jib to build and upload the Docker image to the local Docker images repository.

4.  Open the terminal in the pet-clinic-concierge microservice root directory:

    a. Run the jib command to build Docker mvn compile jib:dockerBuild.

    b. Wait for jib to build and upload the Docker image to the local Docker images repository.

5.  Open any Bash terminal and change the directory to where you have checked out the pet-clinic docker-compose.yml file:

    a. Run docker compose up -d.

    b. Wait for Docker to finish booting up the pet-clinic stack.

Once the `pet-clinic` application is instantiated and running in Docker, we need to configure Kibana to index and show the logs. To index logs in Kibana, perform the following steps:

1. Navigate to Kibana at `http://localhost:5601` and log in using Elasticsearch credentials as mentioned in the `.env` file in the `docker-elk` directory.

2. Open the home page and click on the **Connect to your Elasticsearch index** hyperlink. After clicking on **Connect to your Elasticsearch index**, Kibana will provide a setup page to connect your index (see the following screenshot):

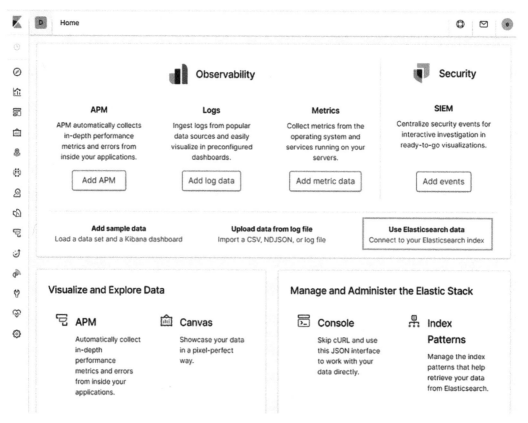

Figure 9.3 – Connecting the Elasticsearch index in Kibana

Kibana provides an intuitive user interface to connect with your Elasticsearch index. Click on the highlighted portion in the screenshot and follow the steps as presented by Kibana.

3. When the setup page loads, enter `logstash` in the **Index Patterns** textbox.

4. Click on the **Next step** button and select **@timestamp** in the configure settings.

5. Then, click on **Create index pattern**.

After a successful index connection, you can go to the **Discover** page and view the application logs as follows:

Figure 9.4 – Viewing the application logs in Discover

On the **Discover** page, we can access the live streaming of logs from various microservices. As we mutated the log events to include `app_name` and `app_port`, we can drill down on both the parameters to see the logs to a specific microservice.

So now, we have implemented an ELK Stack distributed logging that intuitively provides a common place to live access the microservices' logs. If there's any fault in any microservice, you can directly access Kibana and view/search the logs. As you add more microservice instances and components to your runtime topology, ELK will simplify the log management.

In the next section, we will dive into distributed tracing and how we can implement distributed tracing in the `pet-clinic` application.

# Distributed tracing in Micronaut microservices

Distributed tracing is the capability of the system to track and observe the execution flow of a request in distributed systems by collecting data as the request furthers from one service component to another. This trace data compiles metrics such as the time taken at each service along with end-to-end execution flow. Time metrics can help to zero down performance issues such as which service component is a bottleneck to the execution flow and why.

A trace is a Gantt chart-like data structure that stores the trace information in spans. Each span will keep a trace for the execution flow in a particular service component. Furthermore, a span can have a reference to parent span and child spans. Refer to the following figure:

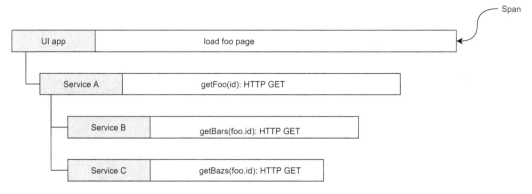

Figure 9.5 – Distributed tracing

In the preceding diagram, we can see the traces/spans for loading the foo page on the user interface app. It first calls **Service A** to get the foo object, which in turn calls **Service B** and **Service C** to get Bars and Bazs for foo, respectively. The time taken for the whole execution will be the cumulative total of execution times at various service components.

In the next section, we will implement a distributed tracing solution in the pet-clinic application.

# Implementing distributed tracing in Micronaut

In order to get hands-on with distributed tracing in Micronaut, we will implement Zipkin-based tracing in the pet-clinic application.

We will run a Zipkin instance in Docker. To run Zipkin in Docker, perform the following steps:

1.  Open any Bash terminal.

2.  Run the docker run -d -p 9411:9411 openzipkin/zipkin command.

    Wait for Docker to download and instantiate Zipkin on port 9411. After successful instantiation, you can verify Zipkin by accessing http://localhost:9411/zipkin.

3.  Next, we will begin with the pet-clinic-concierge service, which is the API gateway. Add the following dependencies to the pet-clinic-concierge POM:

```
<!-- Distributed tracing -->
<dependency>
    <groupId>io.micronaut</groupId>
    <artifactId>micronaut-tracing</artifactId>
    <version>${micronaut.version}</version>
    <scope>compile</scope>
</dependency>
<dependency>
    <groupId>io.zipkin.brave</groupId>
    <artifactId>brave-instrumentation-http</artifactId>
    <version>5.12.3</version>
    <scope>runtime</scope>
</dependency>
<dependency>
```

```
<groupId>io.zipkin.reporter2</groupId>
<artifactId>zipkin-reporter</artifactId>
<version>2.15.0</version>
<scope>runtime</scope>
</dependency>
<dependency>
<groupId>io.opentracing.brave</groupId>
<artifactId>brave-opentracing</artifactId>
<version>0.37.2</version>
<scope>compile</scope>
</dependency>
```

By importing the preceding dependencies, we can leverage Micronaut as well as third-party code artifacts for distributed tracing.

4.  To enable the distributed tracing, we also need to amend `application.properties`. Add the following properties related to Zipkin:

```
tracing:
  zipkin:
    http:
      url: http://host.docker.internal:9411
    enabled: true
    sampler:
      probability: 1
```

The preceding application properties for Zipkin are added at the root level. In `url`, we specified a Docker instance of Zipkin running on localhost. Furthermore, in `sampler.probability`, we specify the value as 1 that will enable the tracing for all the user requests. This probability can be reduced to any value between 0 and 1 wherein 0 means never sample and 1 means sample every request.

5.  Next, we need to tag the controller methods for spans. For managing the spans, we have the following two tags in Micronaut:

    a. `@NewSpan`: This will create a new span beginning from the method it's tagged on.

    b. `@ContinueSpan`: This will continue the previous span.

Since all the client controllers in `pet-clinic-concierge` are the interfacing points to any upstream consumers, we will use `@NewSpan` on these methods so that a new trace can begin. The following are the span-related changes in `OwnerResourceClientController`:

```
@Controller("/api")
public class OwnerResourceClientController {
    @Inject
    OwnerResourceClient;
    @NewSpan
    @Post("/owners")
    public HttpResponse<OwnerDTO>
      createOwner(@SpanTag("owner.dto") OwnerDTO ownerDTO) {
          return ownerResourceClient.createOwner(ownerDTO);
    }
    @NewSpan
    @Put("/owners")
    HttpResponse<OwnerDTO> updateOwner
      (@SpanTag("owner.dto") @Body OwnerDTO ownerDTO) {
          return ownerResourceClient.updateOwner(ownerDTO);
    }
    ...
}
```

Similar changes to annotate client controller methods should be made in all the other clients for the `pet-owner`, `pet-clinic`, and `pet-clinic-reviews` microservices.

Next, we need to modify the `pet-clinic` microservice for distributed tracing.

## Modifying the pet-clinic microservice for distributed tracing

Continuing with distributed tracing changes, we need to make the required amends in the `pet-owner`, `pet-clinic`, and `pet-clinic-reviews` microservices project POM and application properties as explained in the previous section.

Furthermore, to continue the tracing, we need to annotate controller methods with `@ContinueSpan` tags. Refer to the following code block:

```
@Post("/owners")
@ExecuteOn(TaskExecutors.IO)
@ContinueSpan
public HttpResponse<OwnerDTO> createOwner(@Body OwnerDTO
ownerDTO) throws URISyntaxException {
    ...
}
```

`@ContinueSpan` must be annotated on all the controller methods in all the microservices (excluding `pet-clinic-concierge`, which is an API gateway). `@ContinueSpan` will continue the span/trace from the previous span/trace. In `pet-clinic-concierge`, we annotate the `createOwner()` method with `@NewSpan`, and in the `pet-owner` microservice, we use `@ContinueSpan`. Using these tags in tandem will trace the end-to-end execution flow.

In the next section, we will verify the end-to-end trace for an HTTP request in the `pet-clinic` application.

## Verifying the distributed tracing in the pet-clinic application

To verify the distributed tracing in the `pet-clinic` application, you must have the `pet-clinic` microservice running. We will fetch a list of owners via the API gateway. For this, perform the following steps:

1. Go to `http://localhost:32584/api/owners` in any browser tab or REST client.

2. Navigate to Zipkin to verify the trace for the preceding HTTP GET call at `http://localhost:9411/zipkin`.

3. Click on the **Run Query** button.

4. Go to the `get /api/owners` request in the returned results and click **Show**.

After successfully performing these steps, you will see the following screen:

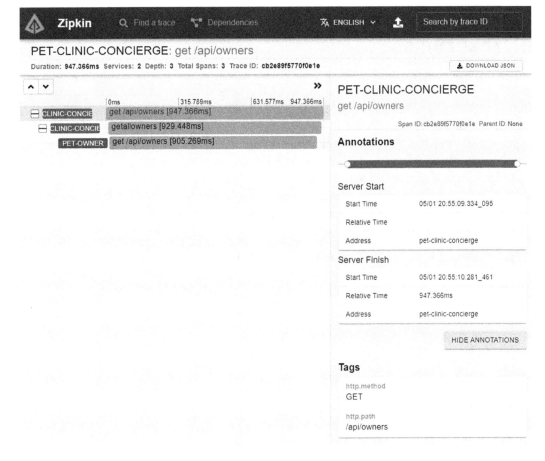

Figure 9.6 – GET owners distributed tracing in Zipkin

Zipkin provides an intuitive user interface for accessing the request execution traces. You can see at first the request reaches pet-clinic-concierge, which is further passed on to the pet-owner microservice. In total it took approximately 948 ms to complete the request with the majority of the time spent on the pet-owner microservice.

In the next section, we will focus on distributed monitoring and how to implement distributed monitoring in the Micronaut framework.

# Distributed monitoring in Micronaut microservices

Monitoring is simply recording the key performance metrics to enhance visibility into the application state. By recording and surfacing the system performance metrics such as CPU usage, thread pools, memory usage, and database connections for all the distributed components, it can provide a holistic picture of how a microservice system is performing at a given point in time. The distributed nature of microservices requires a shift in how the system is monitored. Instead of relying on the host environment monitoring tools, we need a unified monitoring solution that can combine performance metrics from various services and present a one-stop interface. In this section, we will explore how to implement such a distributed monitoring solution for the `pet-clinic` application.

To implement distributed monitoring, we will use a very popular stack of Prometheus and Grafana. Let's look at our system components for distributed monitoring:

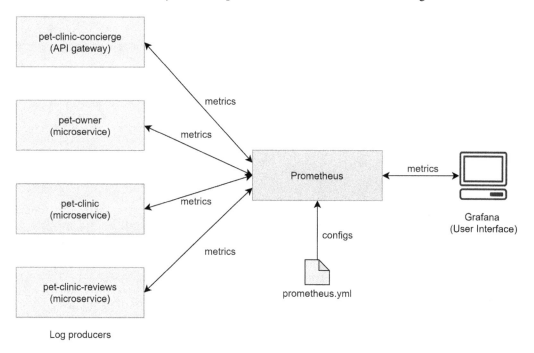

Figure 9.7 – Distributed monitoring using Prometheus and Grafana

As shown in the preceding diagram, the `pet-clinic` microservice will be communicating the metrics to the **Prometheus** server and **Grafana** will get the metrics to present the user interface. Prometheus configurations will be stored in a YAML file.

In the next section, we will begin with setting up Prometheus and Grafana in Docker.

## Setting up Prometheus and Grafana in Docker

Before we instantiate Prometheus and Grafana in Docker, we need to define configurations for Prometheus so that it can pull required metrics from the `pet-clinic` microservice. You can check out `docker-prometheus docker-compose` and `prometheus.yml` from `https://github.com/PacktPublishing/Building-Microservices-with-Micronaut/tree/master/Chapter09/micronaut-petclinic/docker-prometheus`.

Once checked out locally, you can review the `prometheus.yml` file to be as follows:

```
global:
  scrape_interval:     15s
  evaluation_interval: 15s
  external_labels:
      monitor: 'pet-clinic-monitor'
scrape_configs:
  - job_name: 'prometheus'
    scrape_interval: 10s
    static_configs:
        - targets: ['host.docker.internal:9090','node-
            exporter:9110']
  - job_name: 'micronaut'
    metrics_path: '/metrics'
    scrape_interval: 10s
    static_configs:
        - targets: ['host.docker.internal:32581', 'host.
docker.internal:32582', 'host.docker.internal:32583', 'host.
docker.internal:32584']
```

In `prometheus.yml`, we mainly need to configure `scrape_configs`. This will be responsible for invoking the microservice endpoints to get the metrics. We can specify the `pet-clinic` microservice in the targets. In addition, you can note that the scrape interval is `10` seconds. This will configure Prometheus to fetch metrics every 10 seconds.

Next, let's set up our distributed monitoring stack in Docker.

To set up Prometheus and Grafana in Docker, follow these instructions:

1.  Check out `docker-prometheus` from `https://github.com/PacktPublishing/Building-Microservices-with-Micronaut/tree/master/Chapter09/micronaut-petclinic/docker-prometheus`.

2.  Open any Bash terminal (we used Git Bash).

3.  Change directory to where you have checked out `docker-prometheus`.

4.  Run `docker compose up -d`.

5.  Wait for Docker to download the images and instantiate the Prometheus app container.

These instructions will boot up the monitoring app in Docker. You can verify the installation by going to the Docker Dashboard and **Containers** / **Apps**.

In the next section, we will explore how we can integrate the `pet-clinic` microservice into Prometheus.

## Configuring microservices for distributed monitoring

To configure the `pet-clinic` microservice for distributed monitoring, we need to update the `project` POM with `Micrometer` dependencies.

Add the following dependencies to the `pet-owner` project POM:

```
<!-- Micrometer -->
<dependency>
   <groupId>io.micronaut.micrometer</groupId>
   <artifactId>micronaut-micrometer-core</artifactId>
</dependency>
<dependency>
```

```xml
    <groupId>io.micronaut.micrometer</groupId>
    <artifactId>micronaut-micrometer-registry-
    prometheus</artifactId>
</dependency>
<dependency>
    <groupId>io.micronaut</groupId>
    <artifactId>micronaut-management</artifactId>
</dependency>
```

By importing the `micronaut-micrometer` dependencies, we can leverage a distributed monitoring toolkit in the `pet-owner` microservice.

To expose service metrics for Prometheus, we need to expose the `metrics` endpoint in all the `pet-clinic` microservices. We will add a new controller called `PrometheusController` to the `com.packtpub.micronaut.web.rest.commons` package as follows:

```java
@RequiresMetrics
@Controller("/metrics")
public class PrometheusController {
    private final PrometheusMeterRegistry;
    @Inject
    public PrometheusController(PrometheusMeterRegistry
     prometheusMeterRegistry) {
        this.prometheusMeterRegistry =
        prometheusMeterRegistry;
    }
    @Get
    @Produces("text/plain")
    public String metrics() {
        return prometheusMeterRegistry.scrape();
    }
}
```

PrometheusController will expose prometheusMeterRegistry.scrape() on the /metrics endpoint.

prometheusMeterRegistry.scrape() will provide service performance metrics as configured in the application.properties file.

We need to configure the application.properties file as follows:

```
micronaut:
    ...
  metrics:
    enabled: true
    export:
      prometheus:
        enabled: true
        step: PT1M
        descriptions: true

endpoints:
  metrics:
    enabled: false
  prometheus:
    enabled: false
```

In application.properties, we are enabling the metrics and exporting the metrics in Prometheus format. Furthermore, since we are providing our custom /metrics endpoint, we are disabling the metrics and prometheus endpoints in the application properties.

Similarly, we need to modify the project POM, add `PrometheusController`, and update the application properties for the `pet-clinic`, `pet-clinic-reviews`, and `pet-clinic-concierge` microservices. Afterward, we need to rebuild the Docker images for all service projects running the `mvn compile jib:dockerBuild` command in the terminal. Once the Docker images are built and uploaded to the local Docker repository, we need to decommission the old `pet-clinic` application in Docker and rerun `docker compose up -d` to re-instantiate the modified `pet-clinic` application.

In the next section, we will verify the distributed monitoring implementation in the `pet-clinic` application.

# Verifying the distributed monitoring in the pet-clinic application

To verify the distributed monitoring in the `pet-clinic` application, you must have the `pet-clinic` and Prometheus applications running in Docker. You need to follow these instructions to verify integration between Prometheus and the `pet-clinic` application:

1. Access the `/metrics` endpoints for all the microservices to verify that services are exposing metrics to Prometheus.

2. Verify the `pet-owner` metrics by accessing `http://localhost:32581/metrics`.

3. Verify the `pet-clinic` metrics by accessing `http://localhost:32582/metrics`.

4. Verify the `pet-clinic-reviews` metrics by accessing `http://localhost:32583/metrics`.

5. Verify the `pet-clinic-concierge` metrics by accessing `http://localhost:32584/metrics`.

6. Navigate to `http://localhost:9090/graph` and check whether you can see the `system_cpu_usage` metric.

After successful completion of the preceding steps, you will see the following screen:

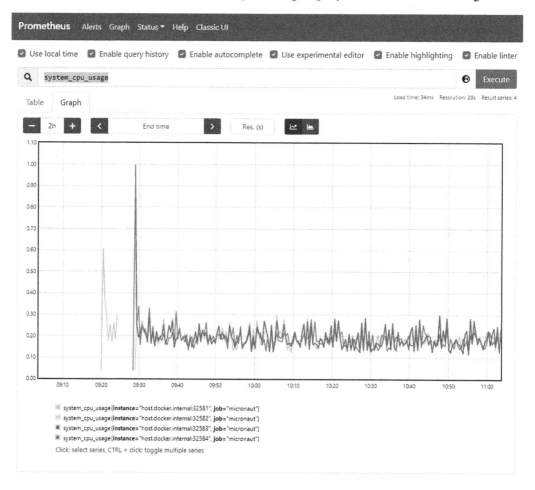

Figure 9.8 – Accessing the system CPU usage graph for the pet-clinic application in Prometheus

In the preceding screenshot, we can verify that the pet-clinic microservice is able to expose the performance metrics on the endpoint and Prometheus can invoke the / metrics endpoints. We can see the system CPU usage graph in Prometheus graphs but as a system admin or developer, you probably need a system dashboard with all the metric graphs in one place.

In the following instructions, we will integrate Grafana with the Prometheus server:

1.  Navigate to http://localhost:3000/ and log in with the username as admin and password as pass.

2.  After logging in, navigate to http://localhost:3000/datasources.

3. Click on the **Add data source** button.

4. Under the **Time series databases** list, select **Prometheus**.

5. In **URL**, provide the value as `http://prometheus:9090`.

   Keep the rest of the values as the defaults.

6. Click on the **Save and test** button. You should get a successful message.

7. Go to the adjacent **Dashboards** tab and click on **Prometheus 2.0 stats**.

After successful completion of these steps, you should see the following dashboard:

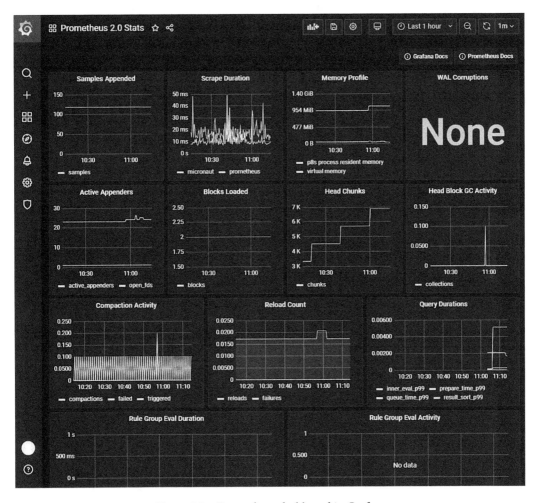

Figure 9.9 – Prometheus dashboard in Grafana

As shown in the preceding screenshot, Grafana provides a very intuitive, unified dashboard for accessing the vital system metrics for all the service components in the `pet-clinic` application. One-stop access to this telemetry data is very handy in addressing any performance issues and system failures in production environments.

In this section, we explored what distributed monitoring is and how we can implement distributed monitoring using Prometheus and Grafana in the Micronaut framework.

# Summary

In this chapter, we began with distributed logging and why it is important in any microservice implementation. We implemented an ELK Stack for distributed logging in the `pet-clinic` application. Furthermore, we dived into using the Kibana user interface for connecting to the Elasticsearch application logs index.

Later, we explored what distributed tracing is and how to implement distributed tracing using Zipkin in the Micronaut framework. We also verified the trace of an HTTP call in the Zipkin user interface.

In the end, we dived into the world of distributed monitoring and implemented a distributed monitoring solution for the `pet-clinic` application using a Prometheus and Grafana stack.

This chapter enhanced your Micronaut microservices journey with the observability patterns that are distributed logging, distributed tracing, and distributed monitoring by enabling you with hands-on knowledge on how to implement these patterns in the Micronaut framework.

In the next chapter, we will explore how to implement an IoT solution in the Micronaut framework.

# Questions

1. What is distributed logging in microservices?

2. How do you run an ELK Stack in Docker?

3. How do you implement distributed logging in the Micronaut framework?

4. How do you connect to a Docker Logstash from the Micronaut microservice?

5. What is distributed tracing in microservices?

6. How do you implement distributed tracing in the Micronaut framework?

7. What is distributed monitoring in microservices?

8. How do you run a Prometheus and Grafana stack in Docker?

9. How do you implement distributed monitoring in the Micronaut framework?

# Section 6: IoT with Micronaut and Closure

This section will wrap up the microservices journey in the Micronaut framework with a booster chapter on the **Internet of Things (IoT)** and later you will learn about building enterprise-grade microservices.

This section has the following chapters:

- *Chapter 10, IoT with Micronaut*
- *Chapter 11, Building Enterprise-Grade Microservices*

# 10

# IoT with Micronaut

**Internet of Things (IoT)** is one of the fastest emerging technologies. It is a network of devices or things. These devices have the same capabilities as sensors or software and can communicate with other devices over the internet. A device or thing can be from various fields and can include things such as light bulbs, door locks, heartbeat monitors, location sensors, and many devices that can be enabled with sensors. It is an ecosystem of smart devices or things with internet capabilities. IoT is trending in various fields. A few of the top trending fields are as follows:

- Home automation
- Manufacturing and industrial applications
- Healthcare and medical science
- Military and defense
- Automotive, transportation, and logistics

Along with these fields, in this chapter, we will learn about the following topics:

- Basics of IoT
- Working with Micronaut Alexa skills

By the end of this chapter, you will be well versed in the preceding aspects concerning IoT with Micronaut integration.

# Technical requirements

All the commands and technical instructions in this chapter can be run on Windows 10 and macOS. The code examples in this chapter are available in this book's GitHub repository at `https://github.com/PacktPublishing/Building-Microservices-with-Micronaut/tree/master/Chapter10/`.

The following tools need to be installed and set up in the development environment:

- **Java SDK**: Version 13 or above (we used Java 14).

- **Maven**: This is optional and only required if you would like to use Maven as the build system. However, we recommend having Maven set up on any development machine. Instructions to download and install Maven can be found at `https://maven.apache.org/download.cgi`.

- **Development IDE**: Based on your preference, any Java-based IDE can be used, but for the purpose of writing this chapter, IntelliJ was used.

- **Git**: Instructions for downloading and installing Git can be found at `https://git-scm.com/downloads`.

- **PostgreSQL**: Instructions for downloading and installing PostgreSQL can be found at `https://www.postgresql.org/download/`.

- **MongoDB**: MongoDB Atlas provides a free online Database-as-a-Service (DBaaS) with up to 512 MB storage. However, if you would prefer to use a local database, then the instructions for downloading and installing MongoDB can be found at `https://docs.mongodb.com/manual/administration/install-community/`. We used a local installation to write this chapter.

- **REST client**: Any HTTP REST client can be used. We used the Advanced REST Client Chrome plugin in this chapter.

- **Docker**: Instructions for downloading and installing Docker can be found at `https://docs.docker.com/get-docker/`.

- **Amazon**: You will need an Amazon account for Alexa, which you can set up at `https://developer.amazon.com/alexa`.

# Basics of IoT

**IoT** is a network of devices or things. These things can be anything – it can be a human wearing a health monitor, a pet wearing a geolocation sensor, a car with a tire pressure sensor, a television with voice/visual capability, or a smart speaker. IoT can also use advanced **machine learning** (**ML**) and **artificial intelligence** (**AI**) capabilities in the cloud to provide next-level services. IoT can make things smart with data collection and automation. The following diagram illustrates IoT:

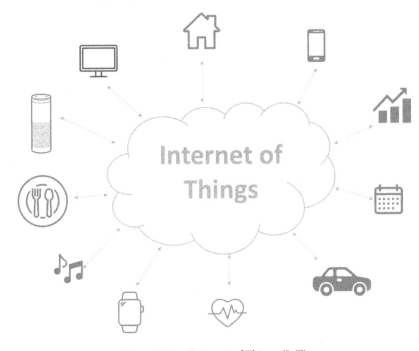

Figure 10.1 – Internet of Things (IoT)

These devices or things have internet capabilities and are interconnected, so they act as an ecosystem. This ecosystem can collect, send, and act based on data it acquires from other things. For example, you can turn on the lights at your home when you arrive.

IoT provides significant benefits to individuals, businesses, and organizations. IoT can reduce manual work and intervention with seamless data transfer between two systems or devices. IoT devices are become more significant every day in the consumer market, be it as locks, doorbells, light bulbs, speakers, televisions, healthcare products, or fitness systems. IoT is mainly accessed now in voice enabled ecosystems such as Google Home, Apple Siri, Amazon Alexa, Microsoft Cortana, Samsung Bixby, and more. There are numerous positive aspects of IoT; however, there are a few cons regarding security and privacy issues.

Now that we have learned about the basics of IoT and its applications, let's understand the basics of Alexa skills.

# Working on the basics of Alexa skills

Alexa is a cloud-based voice recognition service available on millions of devices from Amazon and third-party device manufacturers, such as televisions, Bluetooth speakers, headphones, automobiles, and so on. You can build interactive voice-based request-response applications using Alexa.

Alexa can be integrated into various applications. Alexa also has screen capabilities for displaying responses visually, and Echo Show is an Alexa speaker with a display screen. The following diagram illustrates the Amazon Alexa architecture:

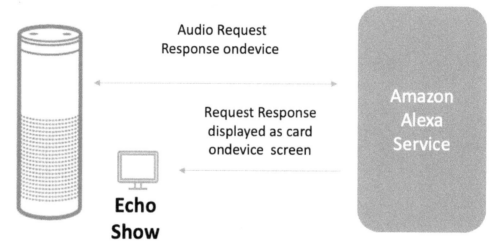

Figure 10.2 – Amazon Alexa architecture

Users can say the wake-up word for the device, which is **Alexa**, and perform an operation. For example, to find the weather in your current location, you can say *Alexa, what is the current weather?* and you will receive a response, such as *The current weather in your location is 28 degrees.* Alexa skills are like apps, and you can enable or disable skills using the Alexa app for a specific device. Skills are voice-based Alexa capabilities.

Alexa can do the following:

- Set an alarm.
- Play music from Spotify, Apple Music, or Google Music.
- Create a to-do list and add items to your shopping list.
- Check the weather.
- Check your calendar.
- Read news briefings.
- Check bank accounts.
- Order in a restaurant.
- Check facts on the internet.

These are a few of the many things that Alexa can perform. Now, let's move on and understand more about Alexa.

## Basics of Alexa skills

Users with any voice-based assistants or tools can use the wake-up word to open the skill or application. For example, with Google Home, we use *Hey Google* or *OK Google*, for Apple Siri, we use *Hey Siri* or *Siri*, and for Amazon Alexa, we use *Alexa*. This wake-up word can be replaced with *Amazon*, *Echo*, or *computer*. All Alexa skills have been designed based on the voice interaction model; that is, phrases you can say to make the skill do something you want, such as *Alexa, turn on the lights* or *Alexa, what is the current temperature?*

Alexa supports the following two types of voice interaction models:

- **Pre-built voice interaction model**: Alexa defines the phrases for each skill for you.
- **Custom voice interaction model**: You define the phrases that the user can say to interact with your skills.

For our working example code, we will use the custom voice interaction model. The following diagram illustrates the process of opening a skill using a custom voice interaction model:

Figure 10.3 – Opening a skill

Now that we know about the wake-up word, the phrase following it is the **launch** word, followed by the **invocation name**. For our sample application, **Pet Clinic**, the launch word will be **open**, followed by the invocation **Pet Clinic**.

The following diagram illustrates the relationship between utterances and intent:

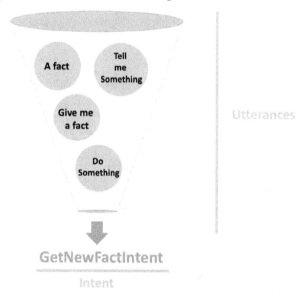

Figure 10.4 – Opening a skill

An utterance is a word users say to Alexa to convey what they want to do such as *Turn on the lights*, *What is the current temperature?*, and so on. Users can say the same thing in different ways, such as *find the temperature, current temperature, outside temperature, the temperature in [location]*, and Alexa will provide pre-build utterances and associated requests as part of the custom voice interaction model. This list of utterances can be mapped to a request or intent.

The following diagram illustrates the custom voice interaction model with a **wake word**, **launch**, **invocation name**, **utterance**, and **intent**:

Figure 10.5 – Opening a skill – Pet Clinic

For our code example, we will use the sequences *Alexa, Open Pet Clinic*, and *Alexa, find nearby Pet Clinics*. Here, the wake-up word is **Alexa**, the launch word is **Open**, and the invocation name is **Pet Clinic**. The utterance can be **find the nearest pet clinic**. We can also have other variations of utterances, such as **find pet clinic**. All these utterances can be mapped to **GetFactByPetClinicIntent**. We will learn about intents in the next section of this chapter.

## Basics of intents

One of the fundamental designs for voice in Alexa is intents. Intents capture events the end user wants to do with voice. Intents represent an action that is triggered by the user's spoken request. Intents in Alexa are specified in a JSON structure called an **intent schema**. The built-in intents include **Cancel**, **Help**, **Stop**, **Navigate Home**, and **Fallback**. Some intents are basic, such as help, and the skills should have a Help Intent.

The following diagram illustrates the built-in intents in the Alexa developer console:

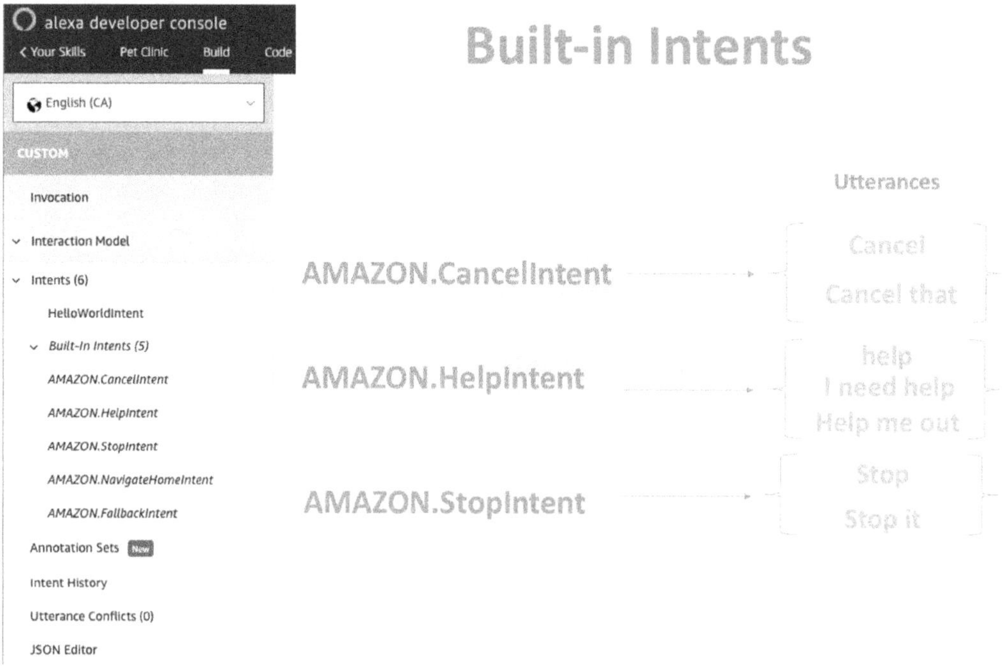

Figure 10.6 – Built-in intents

If we have a website for logging in that has username and password fields and a submit button, there will be a submit intent in the Alexa skill world. However, one big difference is that users can say the submit in different ways; for example, *Submit, submit it, confirm, ok, get, continue*, and so on. These different ways of saying the same thing are called **utterances**. Each intent should include a list of utterances; that is, all the things a user might say to invoke these intents. Intents can have arguments called **slots**, which will not be discussed in this chapter.

Now that we have learned about the basics of Alexa skills by covering utterances, intents, and built-in intents, let's create our first functional Alexa skill.

## Your first HelloWorld Alexa skill

To start creating our Alexa skill, we must navigate to `https://developer.amazon.com/`, select **Amazon Alexa**, and click **Create Alexa Skills**. This will open the Alexa developer console. If you don't have an Amazon developer account, you can create one for free. The following screenshot illustrates the **Create Skill** screen:

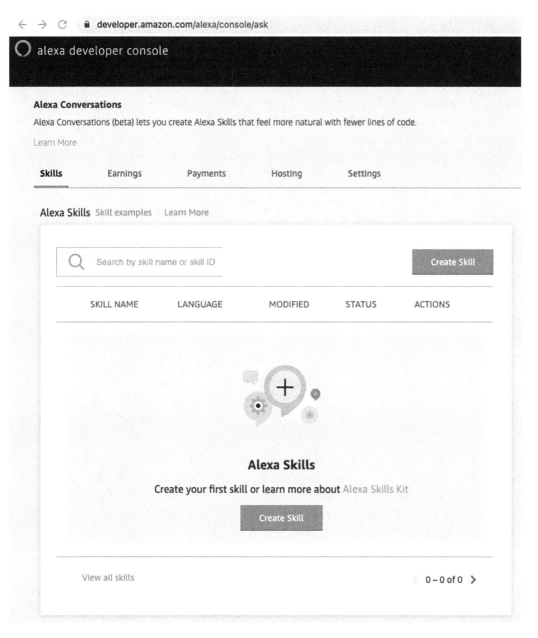

Figure 10.7 – Create Skill screen

In the preceding screenshot, you can see how to create a new skill name called `Pet Clinic`, choose a model to add to your skill option called `Custom`, and choose a method to host your skill's backend resources called `Provision your own`. Choose a template to add to your skill called `Start from Scratch`.

By using the custom voice interaction model, we have learned that we need to create and configure our wake word, launch, invocation name, utterances, and intent. The wake word is configured for the device and is the same for all the skills, so we don't need to change it. In our configuration, we will configure the code launch, invocation, utterances, and intent. The following diagram illustrates the basics of developing Alexa skills:

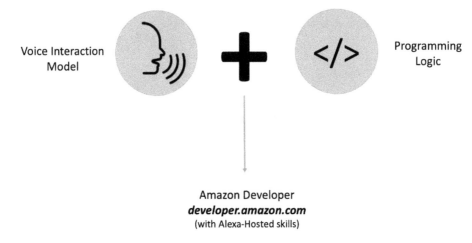

Figure 10.8 – Alexa skills

Alexa skills are based on the **voice interaction model** and **programming logic**. Programming logic can be created using Node.js, Java, Python, C#, or Go. This programming logic allows us to connect to web services, microservices, APIs, and interfaces. With this, you can invoke an internet-accessible endpoint for Alexa skills. The following diagram illustrates the skills developer console:

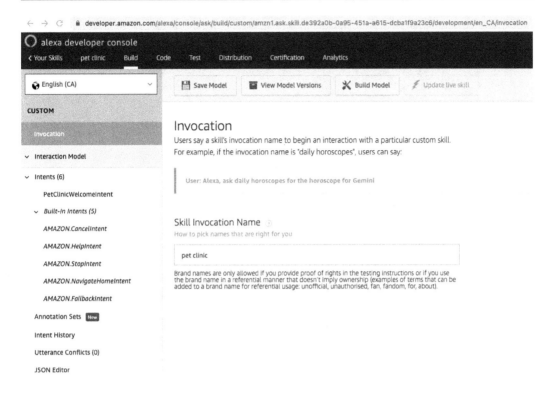

Figure 10.9 – Alexa skills

We can set **Skill Invocation Name** to pet clinic and save it. You can also select the
HelloWorldIntent intent and rename it PetClinicWelcomeIntent. There will
be sample utterances listed in the intent that you can modify manually or use the JSON
Editor and copy the alexa_petclinic_intent_schema.json code from this
book's GitHub repository. The following code illustrates the JSON schema for the intent:

```
{
  "interactionModel": {
    "languageModel": {
      "invocationName": "pet clinic",
      "intents": [
        {
          "name": "AMAZON.CancelIntent",
          "samples": [
            "cancel"
          ]
          ..........................
```

```json
    {
        "name": "PetClinicWelcomeIntent",
        "slots": [],
        "samples": [
          "find near by pet clinics",
          "find pet clinics"
        ]
      }
    ],
    "types": []
  }
 }
}
```

You can configure the intent's invocation name and sample utterance using the JSON configuration file.

Once you have copied the JSON file to the Alexa developer console's JSON editor, click **Save Model** and then **Build Model and Evaluate Model**.

---

**Note**

The preceding configuration is a sample from this chapter's folder on GitHub. The actual schema can be copied from GitHub.

---

Once you have built the model, click **Test** in the Alexa developer console and enable the skill testing process. Now, we need to develop our backend code for the response.

Create a Maven Java project using your favorite IDE. The following dependencies are required for this:

```xml
<dependencies>
    <dependency>
        <groupId>com.amazon.alexa</groupId>
        <artifactId>ask-sdk</artifactId>
        <version>2.20.2</version>
    </dependency>
</dependencies>
```

We will be using Amazon's `ask-sdk` for our backend Java development. You can also configure the dependencies using Gradle. A sample Gradle configuration can be seen in the following code:

```
dependencies {
    compile 'com.amazon.alexa:alexa-skills-kit:1.1.2'
}
```

We need to create a Java class for all the intents. In our JSON schema, we have defined the `CancelIntent`, `HelpIntent`, `StopIntent`, `NavigateHomeIntent`, `FallbackIntent`, and `PetClinicWelcomeIntent` intents. For every intent, we need to create a handler; for example, `PetClinicWelcomeIntent` should have `PetClicWelcomeIntentHandler`. The handler's name will be added to the end of each intent name. We must also create one additional handler that hasn't been configured in the JSON schema, and this is called `LaunchRequestHandler`. This is the first intent that is triggered whenever their skill is launched. The following code illustrates `LaunchRequestHandler`:

```
public class LaunchRequestHandler implements RequestHandler {
    @Override
    public boolean canHandle(HandlerInput handlerInput) {
        return handlerInput.matches
        (requestType(LaunchRequest.class));
    }

    @Override
    public Optional<Response> handle(HandlerInput
      handlerInput) {
        String speechText = "Welcome to Pet Clinic, You can
          say find near by Pet Clinics";
        return handlerInput.getResponseBuilder()
                .withSpeech(speechText)
                .withSimpleCard("PetClinic", speechText)
                .withReprompt(speechText)
                .build();
    }
}
```

LaunchRequestHandler will override the handler method and the response voice message when the skill is launched. This is defined in the code block. In the code, we have a speech text response of *Welcome to Pet Clinic, you can say find near by Pet Clinics*, along with the title of PetClinic.

Now that we have created the handlers (CancelandStopIntentHandler, HelpIntentHandler, LaunchRequestHandler, PetClinicWelcomeIntentHandler, and SessionEndedRequestHandler), we need to create StreamHandler. StreamHandler is the entry point for the AWS Lambda function. All requests that are sent by the end user to Alexa, which invokes your skill, will pass through this class. You need to configure the copy of the skill ID from the Amazon Alexa developer console in the endpoint. Refer to the following code:

```java
public class PetClinicStreamHandler extends SkillStreamHandler
{

    private static Skill getSkill() {
        return Skills.standard()
            .addRequestHandlers(
                    new CancelandStopIntentHandler(),
                    new
                        PetClinicWelcomeIntentHandler(),
                    new HelpIntentHandler(),
                    new LaunchRequestHandler(),
                    new SessionEndedRequestHandler())
                .withSkillId("amzn1.ask.skill.de392a0b-
                0a95-451a-a615-dcba1f9a42c6")
                .build();
    }
    public PetClinicStreamHandler() {
        super(getSkill());
    }
}
```

With that, we have learned about how to use stream handlers and how to invoke intent handlers. Let's learn about the use of the skill ID, which is where you can get information about the skill ID. The following screenshot illustrates the skill ID's location in the developer console:

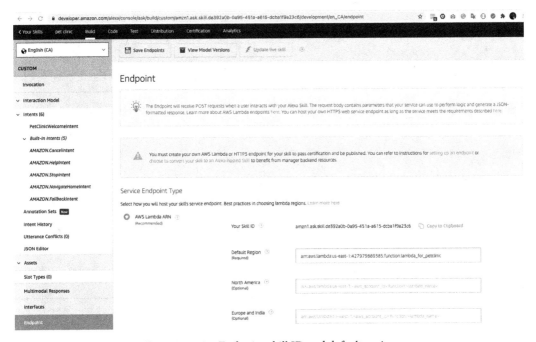

Figure 10.10 – Endpoint skill ID and default region

The skill ID can be found on the Alexa developer console's **Endpoint** screen. The next step is to create the `.jar` file with dependencies for the code. You can execute the `mvn assembly:assembly -DdescriptorId=jar-with-dependencies package` command to create the `.jar` file. This `.jar` file will be located in the target directory, as illustrated in the following screenshot:

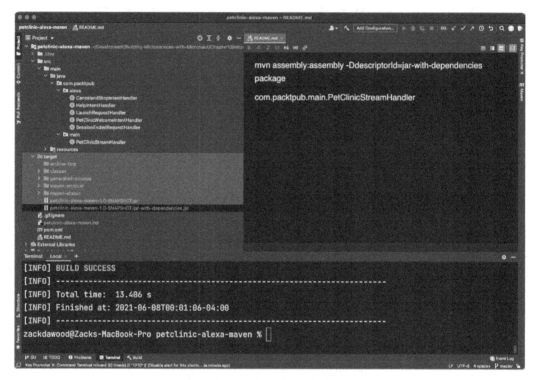

Figure 10.11 – Maven JAR file location

The next step is to create the Amazon Lambda function, which is our backend service code.

Navigate to `https://console.aws.amazon.com/lambda/` to create a function. Name the function `lambda_for_petclinic`, set options to `Author from scratch`, and set runtime to `Java 11`. The user interface is illustrated in the following screenshot:

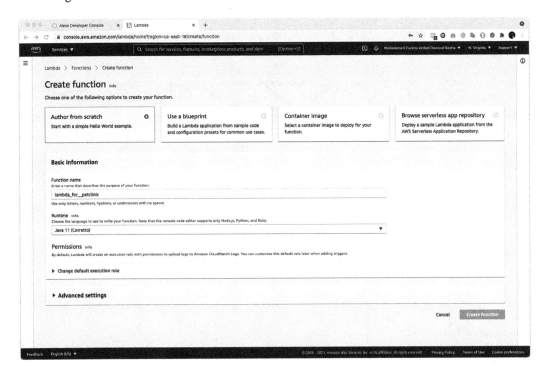

Figure 10.12 – Alexa Create function screen

The next step is to create the trigger with **Trigger configuration** set to **Alexa Skills Kit**, as shown in the following screenshot. You need to copy the Alexa **Skill Id** from the Alexa developer console's **Endpoint** screen. Also, you need to copy the **Function ARN (Amazon Resource Name)** property from the Lambda developer console to the Alexa skill developer console's **Endpoint** screen. The following screenshot illustrates the location of the AWS Lambda function's ARN:

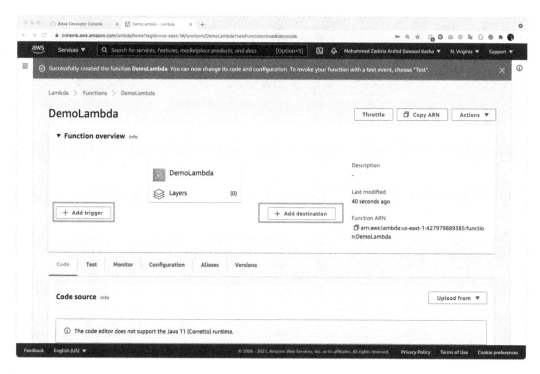

Figure 10.13 – AWS Function ARN\

The **Function ARN** value must be copied from the Alexa developer console **Endpoint** screen to the **Default Region** section or to the location-specific regions, as shown in *Figure 10.10*. The skill ID shown in *Figure 10.10* should be copied to the AWS Lambda trigger screen, as illustrated here:

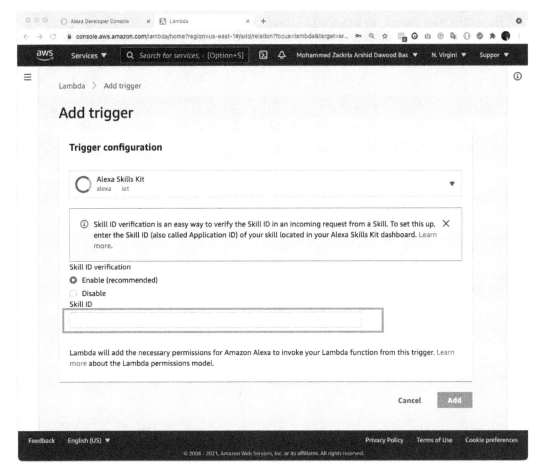

Figure 10.14 – AWS Lambda – Add trigger

Once the trigger has been added with the necessary skill ID and the **Function ARN** ID has been copied to the skill endpoint, click **Save Endpoints**. The next step is to upload the `.jar` file, along with any dependencies (`petclinic-Alexa-maven-1.0-SNAPSHOT-jar-with-dependencies.jar`), to the Lambda function.

The following screenshot illustrates the process of uploading the `.jar` file to the Amazon Lambda function:

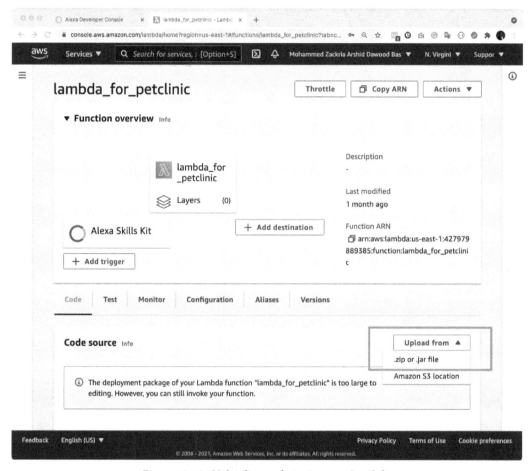

Figure 10.15 –Uploading code to Amazon Lambda

Now that we have created our first Alexa skill and uploaded the necessary code, let's test it out.

# Testing your code

The last and final process is to test the code with the Amazon Alexa Simulator, which is located in the developer console. The following screenshot illustrates how to request the Alexa Simulator:

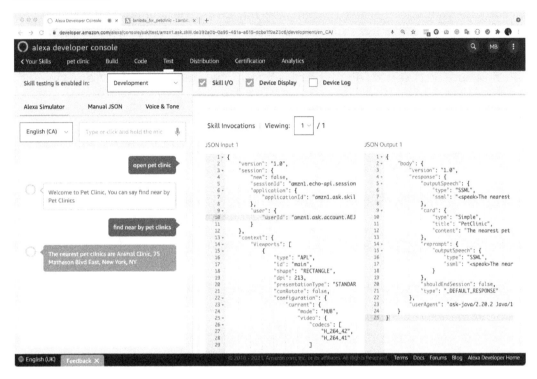

Figure 10.16 – Alexa Simulator request/response

The request/response testing screen accepts text or speech. You can type or say `open pet clinic` and `find nearby pet clinics` here. You should be able to see the response from the Java code in the **JSON Output 1** section. Once you can see the response, this means
we have successfully created our first basic IoT pet clinic example with a request and a response.

We will integrate Micronaut and Alexa in the next section. You can find the complete working example for the `petclinic-alexa-maven` project in this book's GitHub repository.

# Integrating Micronaut with Alexa

As we discussed in the preceding section, in this section, we will start to understand how to integrate Micronaut with Alexa. Micronaut provides various extensions that support **Amazon Web Services (AWS)**. The primary focus here will be on AWS Lambda. The `micronaut-function-aws-alexa` module includes support for building Alexa skills with Micronaut. Micronaut Alexa support can wire up your Alexa application with **AlexaFunction** and supports dependency injection for the following types:

- `com.amazon.ask.dispatcher.request.handler.RequestHandler`
- `com.amazon.ask.dispatcher.request.interceptor.` `RequestInterceptor`
- `com.amazon.ask.dispatcher.exception.ExceptionHandler`
- `com.amazon.ask.builder.SkillBuilder`

Micronaut's `aws-alexa` module simplifies how we can develop Alexa skills with Java, Kotlin, or Groovy. The following code is the Java Maven dependency for the `aws-alexa` module:

```
<dependency>
    <groupId>io.micronaut.aws</groupId>
    <artifactId>micronaut-aws-alexa</artifactId>
</dependency>
```

As we learned in the previous chapters, Micronaut uses Java annotations. To change any Alexa Java handler so that it can work with Micronaut, all we need to do is add the necessary `@Singleton` annotation; that is, `javax.inject.Singleton`. A sample `LaunchRequestHandler` with the `Singleton` annotation is as follows:

```
@Singleton
public class LaunchRequestHandler implements RequestHandler {
    @Override
    public boolean canHandle(HandlerInput handlerInput) {
        return handlerInput.matches
            (requestType(LaunchRequest.class));
    }

    @Override
    public Optional<Response> handle(HandlerInput
     handlerInput) {
```

```
        String speechText = "Welcome to Pet Clinic, You can
          say find near by Pet Clinics";
        return handlerInput.getResponseBuilder()
                .withSpeech(speechText)
                .withSimpleCard("PetClinic", speechText)
                .withReprompt(speechText)
                .build();
    }
}
```

With the help of Micronaut, you can perform unit testing for your intents easily. This is because the @MicronautTest annotation provides seamless unit testing capabilities. Here, we can inject the handler into the unit test cases. The Micronaut framework leverages the Amazon LaunchRequest class to do the following:

```
@MicronautTest
public class LaunchRequestIntentHandlerTest {

    @Inject
    LaunchRequestHandler handler;

    @Test
    void testLaunchRequestIntentHandler() {
        LaunchRequest request =
          LaunchRequest.builder().build();
        HandlerInput input = HandlerInput.builder()
                .withRequestEnvelope
                    (RequestEnvelope.builder()
                        .withRequest(request)
                        .build()
              ).build();

        assertTrue(handler.canHandle(input));
        Optional<Response> responseOptional =
          handler.handle(input);
        assertTrue(responseOptional.isPresent());
        Response = responseOptional.get();
```

```
    assertTrue(response.getOutputSpeech() instanceof
     SsmlOutputSpeech);
    String speechText = "Welcome to Pet Clinic, You can
     say find near by Pet Clinics";
    String expectedSsml = "<speak>" + speechText +
       "</speak>";
    assertEquals(expectedSsml, ((SsmlOutputSpeech)
     response.getOutputSpeech()).getSsml());
    assertNotNull(response.getReprompt());
    assertNotNull(response.getReprompt()
       .getOutputSpeech());
    assertTrue(response.getReprompt().getOutputSpeech()
     instanceof SsmlOutputSpeech);
    assertEquals(expectedSsml,((SsmlOutputSpeech)
       response.getReprompt().getOutputSpeech())
       .getSsml());
    assertTrue(response.getCard() instanceof
       SimpleCard);
    assertEquals("PetClinic", ((SimpleCard)
       response.getCard()).getTitle());
    assertEquals(speechText, ((SimpleCard)
     response.getCard()).getContent());
    assertFalse(response.getShouldEndSession());
}
```

You can find the complete working example for the `petclinic-alexa-micronaut-maven` project in this book's GitHub repository. You can connect to a web service or to a backend database in the handler to send a request and receive a response. The following diagram illustrates the design for Alexa skill integration with the backend:

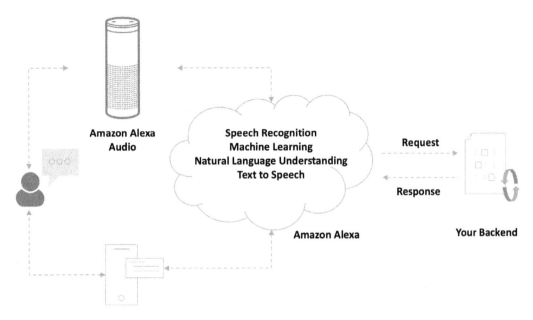

Figure 10.17 – Alexa skills with a custom backend

Your **backend** logic can be embedded in the intent handler. For example, to find the pet clinic's location, you can invoke a web service and add it as a response to speechText, as shown in the following code snippet:

```
@Override
public Optional<Response> handle(HandlerInput
    handlerInput) {
    speechText =  service.getPetClinicLocation();
    return handlerInput.getResponseBuilder()
            .withSpeech(speechText)
            .withSimpleCard("PetClinic", speechText)
            .withReprompt(speechText)
            .build();
}
```

The `speechText` handle method can be added from a microservice call and can retrieve information from a database or service.

Now that we have integrated Micronaut with Alexa, we can control IoT devices with the Voice and Micronaut microservices.

# Summary

In this chapter, we explored how to use the basics of IoT and Amazon Alexa. Then, we dived into creating a Micronaut microservice and integrating it with Amazon Alexa. With this integration, we can control IoT devices with the Voice and Micronaut microservices.

This chapter enhanced your Micronaut microservices journey in IoT. It has equipped you with first-hand knowledge of IoT and Amazon Alexa. Micronaut also supports **Speech Synthesis Markup Language** (**SSML**) and Flash Briefings.

In the next chapter, we will bring all the topics that we have learned about together and take things to the next level by architecting enterprise microservices, looking at OpenAPI, scaling Micronaut, and deep diving into building enterprise-grade microservices.

# Questions

1.  What is IoT?
2.  Name a few devices that are IoT devices.
3.  What is an Alexa skill?
4.  What are Alexa intents?
5.  Which programming languages does Alexa support?
6.  What is the default launch handler class's name?
7.  What is the one change you will need to make to your annotate handlers so that they're compatible with Micronaut?

# 11
# Building Enterprise-Grade Microservices

We have reached the last stage of learning about hands-on microservices with Micronaut. Our journey is now in its final stage; we gained a lot of knowledge of being hands-on in Micronaut in the previous chapters. Now, we must connect all the pieces to build our enterprise-grade microservices with Micronaut. As we already know from our previous chapters, the following are some of the benefits of using Micronaut:

- Modern JVM-based full-stack framework
- Easily testable microservices
- Built for serverless applications
- Reactive stack
- Minimal memory footprint and startup time
- Cloud-native framework
- Multi-language support (Java, Groovy, Kotlin)

In this chapter, we will cover the following topics:

- Bringing it all together
- Architecting enterprise microservices
- Understanding Micronaut's OpenAPI
- Implementing Micronaut's microservices

By the end of this chapter, you will be well-versed in building, architecting, and scaling enterprise-grade microservices.

# Technical requirements

All the commands and technical instructions in this chapter can be run on Windows 10 and Mac OS X. The code examples in this chapter are available in this book's GitHub repository at `https://github.com/PacktPublishing/Building-Microservices-with-Micronaut/tree/master/Chapter11/micronaut-petclinic`.

The following tools need to be installed and set up in your development environment:

- **Java SDK**: Version 8 or above (we used Java 13).
- **Maven**: This is optional and only required if you would like to use Maven as your build system. However, we recommend having Maven set up on any development machine. The instructions for downloading and installing Maven can be found at `https://maven.apache.org/download.cgi`.
- **Development IDE**: Based on your preference, any Java-based IDE can be used, but in this chapter, IntelliJ was used.
- **Git**: The instructions for downloading and installing Git can be found at `https://git-scm.com/downloads`.
- **PostgreSQL**: The instructions for downloading and installing PostgreSQL can be found at `https://www.postgresql.org/download/`.
- **MongoDB**: MongoDB Atlas provides a free online Database-as-a-Service and up to 512 MB of storage. However, if you would prefer to use a local database, then the instructions for downloading and installing MongoDB can be found at `https://docs.mongodb.com/manual/administration/install-community/`. We used a local installation while writing this chapter.

- **REST client**: Any HTTP REST client can be used. We used the Advanced REST Client Chrome plugin in this chapter.

- **Docker**: The instructions for downloading and installing Docker can be found at `https://docs.docker.com/get-docker/`.

- **Amazon**: An Amazon account for Alexa: `https://developer.amazon.com/alexa`.

# Bringing it all together

Let's recap everything we learned about in all the chapters so far. In *Chapter 1*, *Getting Started with Microservices Using the Micronaut Framework*, we started by looking at microservices while using the Micronaut framework. There, we learned about microservices and their evolution: microservices design patterns. We learned about why Micronaut is the best choice for developing microservices and created our first Micronaut application. In *Chapter 2*, *Working on Data Access*, we learned about working on data access.

We started our first pet clinic, pet owner, and pet clinic review Micronaut project. We learned about integrating the persistence layer using the Micronaut framework, as well as about integrating with a relational database using an object-relational mapping Hibernate framework. We created our Micronaut backend database in PostgreSQL, defined relationships among entities, mapped the relationship between entities, and created data access repositories. We also created basic CRUD operations by inserting/creating, reading/fetching, updating, and deleting in the database using Micronaut. After, we integrated the relational database using the MyBatis framework. We also explored NoSQL database functionalities with MongoDB and Micronaut.

In *Chapter 3*, *Working on RESTful Web Services*, we learned about working on RESTful web services using Micronaut. We added RESTful web service capabilities to our pet clinic, pet owner, and pet clinic review Micronaut projects. We learned about data transfer objects, endpoint payloads, map structs, RESTful endpoints, the HTTP server API, validating data, handling errors, versioning the APIs, and the HTTP client API. We performed a RESTful operation on the service with GET, POST, PUT, and DELETE. In *Chapter 4*, *Securing Microservices*, we learned about securing microservices. We created a working example of a RESTful microservice by enabling security aspects. We learned about the basics of Micronaut security by using session authentication for securing the service endpoints, implementing a basic authentication provider, configuring authorizations, granting unauthorized and secure access, using JWT authentication, setting up Keycloak in Docker, using OAuth, setting up the Okta identity provider, and enabling SSL in the Micronaut framework.

In *Chapter 5*, *Integrating Microservices Using the Event-Driven Architecture*, we learned about integrating microservices using the event-driven architecture. We learned about the basics of the event-driven architecture, event streaming with Apache Kafka, and implementing an event producer and event consumer client in the pet clinic reviews microservice. In *Chapter 6*, *Testing Microservices*, we mastered testing our microservices with Micronaut. We learned about the basics of unit testing in the Micronaut framework with JUnit 5, mock testing, service testing, the test suite that's available, and integration testing using test Docker containers.

In *Chapter 7*, *Handling Microservices Concerns*, we learned about handling microservices concerns. We learned about externalizing application configurations, distributed configuration management, documenting the Service API using Swagger, implementing service discovery, creating service discovery using Consul, implementing the API gateway service, and implementing fault tolerance mechanisms using circuit breakers and fallbacks. In *Chapter 8*, *Deploying Microservices*, we learned about deploying microservices. We learned about building container artifacts, containerizing using Jib, deploying container artifacts, using `docker-compose`, and deploying a multi-service application.

In *Chapter 9*, *Distributed Logging, Tracing, and Monitoring*, we learned about distributed logging, tracing, and monitoring. We also learned about the log producer, dispatcher, storage, and visualizer. We implemented Elasticsearch, Logstash, and Kibana. After, we set up ELK in Docker, integrated with some Micronaut microservices, and implemented distributed tracing in Micronaut. Finally, we set up Prometheus and Grafana in Docker. In *Chapter 10*, *IoT with Micronaut*, we learned about IoT with Micronaut. We learned about IoT, Alexa skills, space facts, utterances, intents, your first Alexa skills, the voice interaction model, integrating Micronaut with Alexa, AWS, AlexaFunction, and testing your AlexaFunction with Micronaut.

To summarize, we learned about all the building blocks that are required to create an enterprise application using databases, web services, containers, deployments, testing, configuration, monitoring, event-driven architecture, security, and IoT. All the working examples we covered are available in this book's GitHub repository.

The following diagram provides a summary of all the chapters in this book:

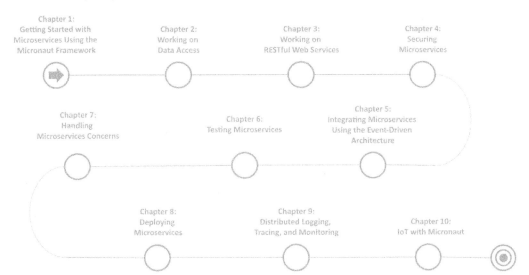

Figure 11.1 – Bringing it all together – chapters roadmap

With that, we have learned about the foundations of creating Micronaut microservices. Now that we are familiar with Micronaut, the necessary development tools, testing, databases, the event-driven architecture, distributed logging, tracing, monitoring, and IoT, we have all the necessary knowledge and skills to create enterprise microservices. We will learn about architecting enterprise microservices in the next section.

# Architecting enterprise microservices

Implementing enterprise microservices requires understanding and motivation from multiple stakeholders in the organization. You need to plan and analyze whether a microservice is the right fit for the problem at hand. If it is a fit, then you must design, develop, deploy, manage, and maintain the service.

Before you start using microservices, let's understand when not to use them. Ask yourself the following questions:

- Does your team know about microservices?

- Is your business mature enough to adopt microservices?

- Do you have an Agile DevOps practice and infrastructure?

- Do you have a scalable on-premises or cloud infrastructure?

- Do you have support to use modern tools and technology?
- Is your database ready to be decentralized?
- Do you have support from all the stakeholders?

If your answer is yes to each of these questions, you can adapt and roll out microservices. The hardest part about rolling out microservices is your data and infrastructure. Traditionally, applications are designed to be big monolithic apps compared to decentralized, loosely coupled microservices. When you architect a microservice, you need to apply multiple techniques during the various phases. The following diagram illustrates the stages of rolling out enterprise microservices:

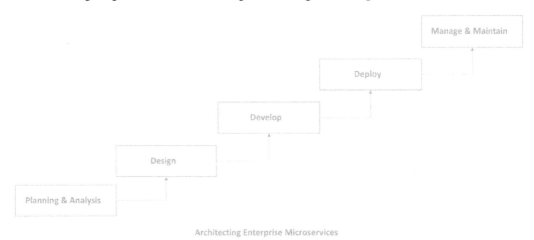

Architecting Enterprise Microservices

Figure 11.2 – Phases of architecting and rolling out enterprise microservices

We'll look at each of these stages in the following sections.

## Planning and analysis

Before you can create an architecture for an enterprise microservice, you need to analyze whether it is a good fit based on your requirements. You can also implement microservices for just a part of the application.

Transitioning from a traditional monolithic architecture to microservices can be a time-consuming and complex process. However, if it is planned well, it can be rolled out seamlessly. Getting all stakeholders to support this is crucial, and this can be accomplished in the planning and analysis phase. Team members having knowledge of microservices is also vital and a critical success factor.

# Design

During the design phase, you can use design patterns, as we learned in *Chapter 1*, *Getting Started with Microservices Using Micronaut Framework*. Design patterns are reusable proven solutions for recurring business or technology problems. The following are the design patterns that we have explored so far in this book:

- Decomposition by business capability

- Decomposing by domains/subdomains

- API gateway pattern

- Chained microservices pattern

- Database per service

- Command query responsibility segregation pattern

- Service discovery pattern

- Circuit breaker pattern

- Log aggregation pattern

Design patterns are continuously evolving. Always check for new patterns in the industry and evaluate whether they are a good fit for your solution. The next important area to consider when designing a microservice is security. We learned about securing microservices in *Chapter 4*, *Securing Microservices*. Here, we learned about evaluating authentication strategies, security rules, privilege-based access, OAuth, and SSL.

Other factors to consider when designing are to check for privacy-specific standards in the application such as **HIPAA** (short for **Health Insurance Portability and Accountability Act**), **General Data Protection Regulation (GDPR)**, **Personal Information Protection and Electronic Documents Act (PIPEDA)**, Bank Act, and so on. Check for encryption standards such as encryption at rest and encryption in transit, whether your data is encrypted before transmission, or whether your data has been stored and encrypted. Check for the security protocol version currently being used and apply the latest stable, supported version patches. The data retention strategy is another area to think about while designing a microservice. How long do you need to store the data for, and what is the archival strategy for data and log files? Also, some countries have regulations where you can store the data, check the requirements, and consider them while designing.

# Develop

Development is an essential phase in implementing microservices. Always use the latest stable, supported version for development. Use the Micronaut framework to perform automated testing. Test at various levels, such as unit testing, service testing, and integration testing. In *Chapter 6, Testing the Microservices*, we learned about testing microservices. Here, we learned that you should always emulate a real-world environment using containers or a cloud infrastructure, as well as using mocking and spying concepts during testing.

If you create a separate version control strategy for each service, the service can be stored in separate repositories with the required configuration and logs. Make sure that you synchronize your development, QA, UAT, and PROD environments and that you have an identical infrastructure across various stages of development. During development, think about backward compatibility for the microservices if you would like to support older versions of the microservices. Have a separate database for each microservice to attain their fullest potential.

# Deploy

When it comes to deployment, you should use automated tools and techniques and leverage rapid application deployment strategies. We learned about deploying microservices in *Chapter 8, Deploying Microservices*. Use tools such as containers, virtual machines, the cloud, Jib, and Jenkins, and use infrastructure efficiently – don't overallocate. Finally, ensure that you have a dedicated microservices DevOps strategy to facilitate **continuous integration** and **continuous delivery** (**CI/CD**).

# Manage and maintain

Maintaining multiple microservices is crucial and complex. You should monitor your application with distributed logging, tracing, and monitoring, as we learned in *Chapter 9, Distributed Logging, Tracing, and Monitoring*. Use tools such as Elasticsearch, Logstash, Kibana, Prometheus, and Grafana to do so. Monitor the code base's health using Sonar DevSecOps and check for security vulnerabilities in the code periodically. You should always update the technology's version, operating system, and tools frequently. Also, monitor your CPU usage, memory footprint, and storage space in real time.

Finally, scale your infrastructure at runtime to avoid hardware overuse. We will discuss how to scale Micronaut in the upcoming sections.

Now that we know how to architect the microservices, let's understand more about Micronaut's OpenAPI.

# Understanding Micronaut's OpenAPI

APIs are generic languages for machines to interact with each other. Having an API definition ensures there is a formalized specification in place. All APIs should have a specification, which improves development efficiency and reduces interaction problems. The specifications act as documentation that helps third-party developers or systems to understand the service easily. Micronaut supports OpenAPI (Swagger) YAML at compile time. We learned about this in *Chapter 7, Handling Microservices Concerns*. The **OpenAPI Initiative (OAI)** was previously known as the **Swagger Specification**. It is used to create machine-readable interface files for describing, producing, consuming, and visualizing RESTful web services. The **OAI** is now a consortium that promotes a vendor-neutral description format. OpenAPI is also called a *public API*, which is publicly made available to software developers and companies. Open APIs can be implemented with REST APIs or SOAP APIs. RESTful APIs are the most popular trending API format used in the industry. OpenAPI must have strong encryption and security in place. These APIs can be public or private (closed). A public OpenAPI can be accessed over the internet; however, a private OpenAPI can only be accessed in the intranet within a firewall or VPN service. Open APIs generate accurate documentation, such as all the required meta-information, reusable components, and endpoint details. There are several versions of the **OpenAPI Specification (OAS)**; the current version is 3.1.

You can learn more about OpenAPI at `https://www.openapis.org/`.

Now that we have learned about Micronaut's OpenAPI, lets, understand scaling Micronaut in the enterprise.

## Scaling Micronaut

When designing and implementing an enterprise application, the ability to scale needs to be planned. One of the biggest advantages of using microservices is scalability. Scaling Micronaut services is a crucial factor by design. It is more than just handling volume – it is about scaling with minimal effort involved. Micronaut makes it easier to identify scaling problems and then resolve challenges at each microservice level. Micronaut microservices are single-purpose applications that can be assembled to build large enterprise-scale software systems. Scaling at runtime is a vital factor in modernizing the enterprise.

There are three types of scaling: $x$-axis, $y$-axis, and $z$-axis scaling. The following diagram illustrates the x-axis (horizontal scaling) and y-axis (vertical scaling):

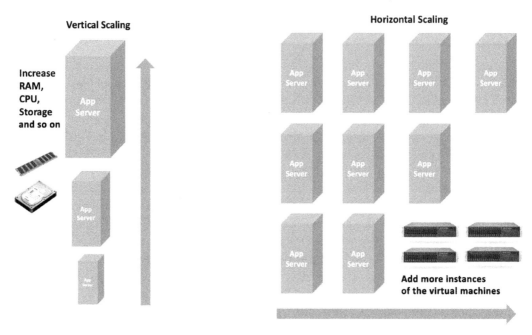

Figure 11.3 – Vertical scaling versus horizontal scaling

Horizontal scaling is also called x-axis scaling. In horizontal scaling, we scale by creating new servers or virtual machines; the entire infrastructure for the service is scaled. For example, if 10 services are running in a virtual machine, we need to add a virtual machine with 10 services if one service requires additional capacity. If you analyze this scenario, there will be unused server capacity as only one service required an additional CPU instead of 10. However, this type of scaling provides unlimited scaling of the infrastructure.

Vertical scaling is also called y-axis scaling. In vertical scaling, we scale by adding capacity to the existing servers. Capacity is scaled by adding additional CPU, storage, and RAM. For example, if 10 services are running in a virtual machine, if one service requires additional capacity such as RAM and CPU, we need to add extra RAM and CPU to the same virtual machine. This is the fundamental difference between horizontal and vertical scaling. However, vertical scaling has a limitation: it cannot scale beyond a specific limit. The following diagram illustrates the x-axis (horizontal scaling) and z-axis (microservices horizontal scaling):

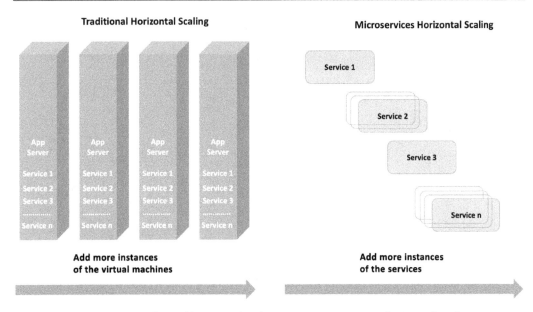

Figure 11.4 – Traditional horizontal scaling versus microservices horizontal scaling

Microservices horizontal scaling is also known as z-axis scaling. This is almost the same as traditional horizontal scaling. For example, if 10 services are running in a microservice container environment, if one service requires additional capacity, we need to add one microservice container environment instead of 10. If you analyze this scenario, you will see that this is the most optimal use of the available capacity. This type of scaling allows you to scale the infrastructure as many times as you like and is the most cost-efficient method. Its performance is a lot better than in a non-scaled environment. You can scale with containers and also with cloud infrastructure. The capabilities of autoscaling are very powerful and come in handy for microservices.

Now we have learned about scaling microservices, let's implement the enterprise microservices with all the features we have learned.

# Implementing Micronaut's microservices

Now, let's implement what we have learned so far in this chapter. You can use the code in this chapter's GitHub repository. We will use the four projects we've covered in this book – pet clinic, pet owner, pet reviews, and concierge. We will also be using a Zipkin container image for distributed tracing, Prometheus for metrics and monitoring, and the `elk` container image for Elasticsearch, Logstash, and Kibana.

The following screenshot illustrates the list of projects in this book's GitHub repository that we will be using:

Figure 11.5 – GitHub projects for our implementation

Follow these steps:

1.  The first step is to set up Keycloak. Please refer to *Chapter 4, Securing Microservices*, the *Setting up Keycloak as the identity provider* and *Creating a client on the Keycloak server* sections.

    The following command can be run to create the Keycloak Docker image:

    ```
    docker run -d --name keycloak -p 8888:8080 -e KEYCLOAK_
    USER=micronaut -e KEYCLOAK_PASSWORD=micronaut123 jboss/
    keycloak
    ```

    > **Note**
    >
    > Setting up Keycloak is an important activity, so please ensure that you refer to *Chapter 4, Securing Microservices*.

    Once your Keycloak image has been created, start the Keycloak Docker image. The following screenshot illustrates Keycloak running in a Docker container on port 8888:

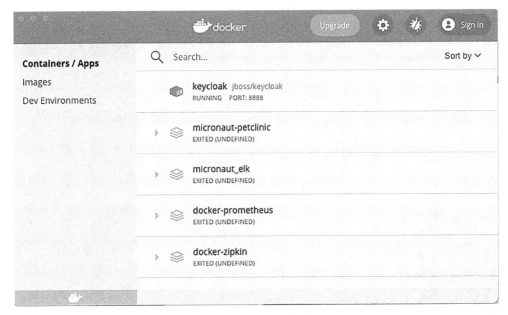

Figure 11.6 – Docker Keycloak running

2.   Once Keycloak is up and running, the client secret key needs to be copied from
     the **Clients | Credentials** screen. Go to **Keycloak | Clients | Credentials | Secret**.

     The following screenshot illustrates the location of your Keycloak **client ID
     and secret**:

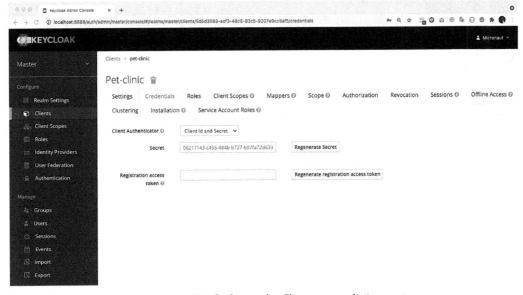

Figure 11.7 – Keycloak portal – Clients – pet-clinic secret

> **Note**
>
> The Keycloak `pet clinic` user and clients should be set up already. Please refer to *Chapter 4, Securing Microservices*, for more information.

3.  Change the timeout settings in Keycloak. Go to the **Keycloak** console | **Clients** | **Settings** | **Advanced Settings**. **15 Minutes** is recommended for testing the sample code.

    The following screen illustrates the Keycloak **Access Token Lifespan** location:

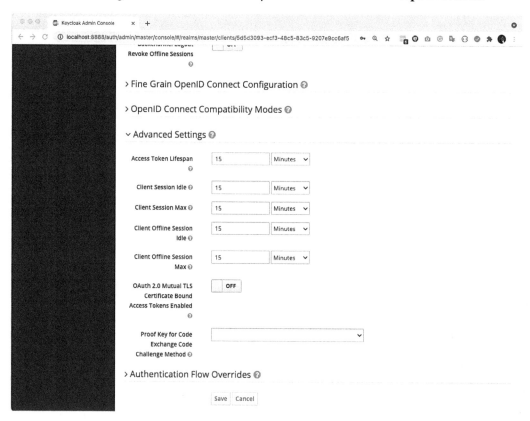

Figure 11.8 – Keycloak Access Token Lifespan

4.  The secret needs to be copied so that it can be replaced in the pet-clinic, pet-owner, pet-clinic-reviews, and pet-clinic-concierge projects. The YAML application configuration file, client-secret, needs to be updated from the Keycloak secret, as shown in the preceding screenshot. The following are the files that must be updated:

    ```
    Chapter11/micronaut-petclinic/pet-clinic-reviews/src/
    main/resources/application.yml
    ```

    ```
    Chapter11/micronaut-petclinic/pet-owner/src/main/
    resources/application.yml
    ```

    ```
    Chapter11/micronaut-petclinic/pet-clinic/src/main/
    resources/application.yml
    ```

    ```
    Chapter11/micronaut-petclinic/pet-clinic-concierge/src/
    main/resources/application.yml
    ```

    The following screenshot illustrates a sample .yaml file configuration where the client secret ID needs to be replaced:

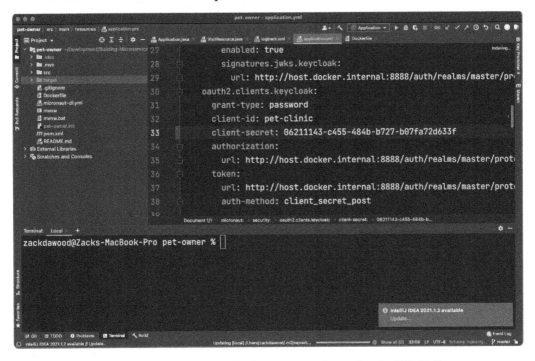

Figure 11.9 – Keycloak portal – client secret in the application YAML file

5.  Once every client secret has been updated, execute the Maven Docker build in all four projects (`pet-clinic-concierge`, `pet-clinic-reviews`, `pet-clinic`, and `pet-owner`). The following Maven command creates a Docker image for each project:

```
mvn clean compile jib:dockerBuild
```

> **Note**
>
> The Postgres database and MongoDB database must be running. Also it must have the database table with data for `pet-clinic` and `pet-reviews`. Refer to *Chapter 02, Working on Data Access* for more details.

6.  Check the Docker setting resources. You need four CPUs and at least 6 GB of memory. Go to the **Docker** settings | **Resources** | **Advanced**:

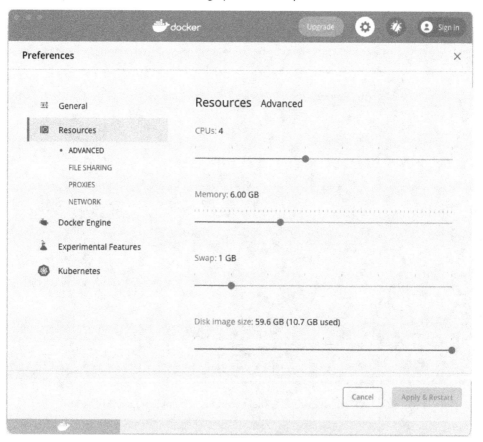

Figure 11.10 – Docker CPU and memory settings

7.  The next step, after creating the Docker images, is to create `docker-compose` for Kafka, Zipkin, Prometheus, and ELK. Execute the following command in a terminal or console to create the Docker image:

```
docker-compose create
```

or

```
docker-compose up -d
```

First, Docker Compose needs to be created for **Kafka**. The configuration file is located here:

```
Chapter11/micronaut-petclinic/pet-clinic-reviews/src/
main/resources/kafka-zookeeper-kafdrop-docker/docker-
compose.yml
```

This is how the file works:

```
zackdawood@Zacks-MacBook-Pro kafka-zookeeper-kafdrop-docker % docker-compose create
WARNING: The create command is deprecated. Use the up command with the --no-start flag
instead.
Creating kafka-zookeeper-kafdrop-docker_zookeeper_1 ... done
Creating kafka-zookeeper-kafdrop-docker_kafka_1      ... done
Creating kafka-zookeeper-kafdrop-docker_kafdrop_1    ... done
zackdawood@Zacks-MacBook-Pro kafka-zookeeper-kafdrop-docker %
```

Figure 11.11 – Docker Compose – Kafka

Next, Docker Compose needs to be created for **Zipkin**. The configuration file is located here:

```
Chapter11/micronaut-petclinic/docker-zipkin/docker-
compose.yml
```

This is how the file works:

```
Last login: Sat Jun  5 23:01:09 on ttys000
zackdawood@Zacks-MacBook-Pro docker-zipkin % docker-compose up -d
Docker Compose is now in the Docker CLI, try `docker compose up`

Creating network "docker-zipkin_default" with the default driver
Creating docker-zipkin_zipkin_1 ... done
zackdawood@Zacks-MacBook-Pro docker-zipkin %
```

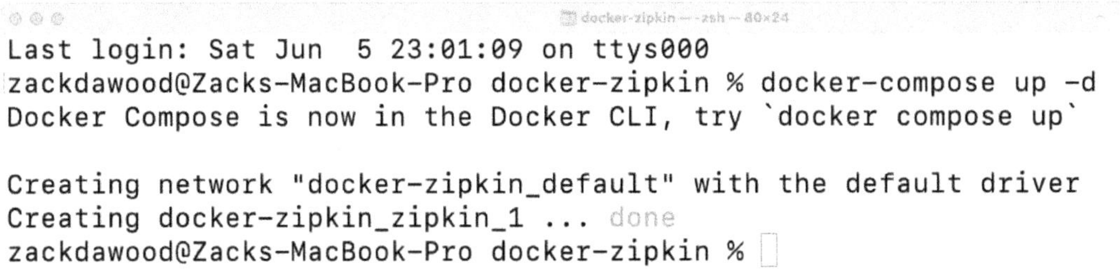

Figure 11.12 – Docker Compose – Zipkin

Next, Docker Compose needs to be created for **Prometheus**. The configuration file is located here:

```
Chapter11/micronaut-petclinic/docker-prometheus/docker-
compose.yml
```

This is how the file works:

```
Last login: Sat Jun  5 23:20:53 on ttys004
zackdawood@Zacks-MacBook-Pro docker-prometheus % docker-compose up -d
Docker Compose is now in the Docker CLI, try `docker compose up`

Creating network "docker-prometheus_default" with the default driver
Creating docker-prometheus_node-exporter_1 ... done
Creating docker-prometheus_prometheus_1    ... done
Creating docker-prometheus_grafana_1       ... done
zackdawood@Zacks-MacBook-Pro docker-prometheus %
```

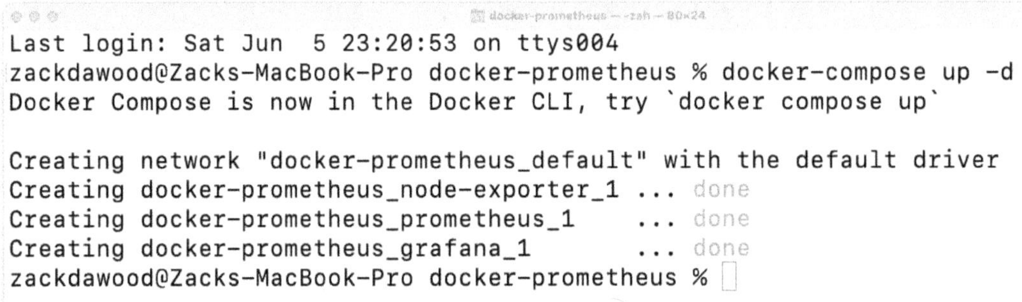

Figure 11.13 – Docker Compose – Prometheus

Now, Docker Compose needs to be created for **ELK**. The configuration file is located here:

```
Chapter11/micronaut-petclinic/docker-prometheus/docker-
elk
```

This is how the file works:

```
Last login: Sat Jun  5 23:23:40 on ttys000
zackdawood@Zacks-MacBook-Pro docker-elk % docker-compose up -d
Docker Compose is now in the Docker CLI, try `docker compose up`

Creating network "micronaut_elk_internal" with the default driver
Creating micronaut_elk_elasticsearch_1 ... done
Creating micronaut_elk_logstash_1       ... done
Creating micronaut_elk_kibana_1         ... done
zackdawood@Zacks-MacBook-Pro docker-elk %
```

Figure 11.14 – Docker Compose – ELK

Finally, Docker Compose needs to be created at the parent level. You can do this by going to the following location:

```
Chapter11/micronaut-petclinic/docker-compose.yml
```

Now, this is how the file works:

```
Last login: Sat Jun  5 23:26:00 on ttys005
zackdawood@Zacks-MacBook-Pro micronaut-petclinic % docker-compose up -d
Docker Compose is now in the Docker CLI, try `docker compose up`

Creating network "micronaut-petclinic_default" with the default driver
Creating micronaut-petclinic_consul_1     ... done
Creating micronaut-petclinic_zookeeper_1 ... done
Creating micronaut-petclinic_kafka_1                ... done
Creating micronaut-petclinic_pet-owner_1            ... done
Creating micronaut-petclinic_pet-clinic-concierge_1 ... done
Creating micronaut-petclinic_pet-clinic-reviews_1   ... done
Creating micronaut-petclinic_pet-clinic_1           ... done
Creating micronaut-petclinic_kafdrop_1              ... done
zackdawood@Zacks-MacBook-Pro micronaut-petclinic %
```

Figure 11.15 – Docker Compose Micronaut pet clinic

Note that the security configuration protects all the projects using Keycloak, except pet-clinic-reviews:

- pet-clinic-concierge: Protected using Keycloak

- pet-clinic: Protected using Keycloak

- pet-owner: Protected using Keycloak

- pet-clinic-reviews: Unprotected

8.  Before testing the applications, check whether all the applications are running in the Docker container. The following screenshot illustrates the application running in a Docker container successfully:

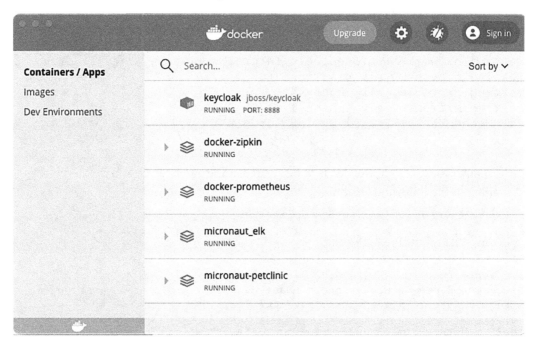

Figure 11.16 – Docker container running all required applications

9.  Now that all the projects are running in Docker, let's test the URL's integration. You can call all the services from the API gateway, as follows:

- pet-owner: http://localhost:32584/api/owners

- pet-clinic: http://localhost:32584/api/vets

- pet-clinic-reviews: http://localhost:32584/api/vet-reviews

    Once you have the apps running in Docker, you need to obtain the security token by invoking the following API on Keycloak:

```
curl -L -X POST 'http://localhost:8888/auth/realms/
master/protocol/openid-connect/token' \
-H 'Content-Type: application/x-www-form-urlencoded' \
--data-urlencode 'client_id=pet-clinic' \
--data-urlencode 'grant_type=password' \
--data-urlencode 'client_secret=PUT_CLIENT_SECRET_HERE' \
```

```
--data-urlencode 'scope=openid' \
--data-urlencode 'username=alice' \
--data-urlencode 'password=alice'
```

> **Note**
>
> PUT_CLIENT_SECRET_HERE must be replaced with the Keycloak credentials secret.

Now, this is what the request in the console looks like:

```
zackdawood@Zacks-MacBook-Pro micronaut-petclinic % curl -L -X POST 'http://local
host:8888/auth/realms/master/protocol/openid-connect/token' \
-H 'Content-Type: application/x-www-form-urlencoded' \
--data-urlencode 'client_id=pet-clinic' \
--data-urlencode 'grant_type=password' \
--data-urlencode 'client_secret=06211143-c455-484b-b727-b07fa72d633f' \
--data-urlencode 'scope=openid' \
--data-urlencode 'username=alice' \
--data-urlencode 'password=alice'
```

Figure 11.17 – curl request to get the JWT

We can then check that the `curl` request returns an access token:

```
zackdawood@Zacks-MacBook-Pro micronaut-petclinic % curl -L -X POST 'http://localhost:8888/auth/realms/master/protocol/openid-connect/token
' \
-H 'Content-Type: application/x-www-form-urlencoded' \
--data-urlencode 'client_id=pet-clinic' \
--data-urlencode 'grant_type=password' \
--data-urlencode 'client_secret=06211143-c455-484b-b727-b07fa72d633f' \
--data-urlencode 'scope=openid' \
--data-urlencode 'username=alice' \
--data-urlencode 'password=alice'
```

{"access_token":"eyJhbGciOiJSUzI1NiIsInR5cCIgOiAiSldUIiwia2lkIiA6ICIweThZU3VsUzYtZlYzWUNudFI2RjFuYUNVYVBUNkplT3k2amR5cUVXNnpnIn0.eyJleHAiO
jE2MjI5NTE0MzUsImlhdCI6MTYyMjk1MDUzNSwianRpIjoiMTBiNDc3MmMtZTgzYi00M2M5LTk5YTktYjhlYTMxYWUyYTRjIiwiaXNzIjoiaHR0cDovL2xvY2FsaG9zdDo4ODg4L2F
1dGgvcmVhbG1zL21hc3RlciIsImF1ZCI6ImFjY291bnQiLCJzdWIiOiJkYTU0ZTdkNy00MGE5LTQyNWEtODM4ZC1iM2U2NTU5ZjVjNzUiLCJ0eXAiOiJCZWFyZXIiLCJhenAiOiJwZ
XQtY2xpbmljIiwic2Vzc2lvbl9zdGF0ZSI6ImE3MDhlZTQ5LTgwNWUtNDQyNy04NWE1LTEyZDM4NjQwY2FkNiIsImFjciI6IjEiLCJyZXNvdXJjZV9hY2Nlc3MiOnsiYWNjb3VudCI
6eyJyb2xlcyI6WyJtYW5hZ2UtYWNjb3VudCIsIm1hbmFnZS1hY2NvdW50LWxpbmtzIiwidmlld1dy1wcm9maWwxlI119fSwic2NvcGUiOiJvcGVuaWQgZW1haWwgcHJvZmlsZSIsImVtY
WlsX3ZlcmlmaWVkIjpmYWxzZSwicm9sZXMiOilsiZGVmYXVsdC1yb2xlcy1tYXN0ZXIiLCJwZXQtc2xpbmljIiwib2ZmbGluZV9hY2Nlc3MiLCJ1bWFfYXV0aG9yaXphdGlvbiJdLXZmM2saW5lX2F2Y2VzcyIsInVtYV9hdXRob3JpemF0aW9u
uIiwicGV0LWNsaW5pY1hZG1pbiJdLCJwcmVmZXJyZWRfdXNlcm5hbWUiOiJhbGljZSJ9.bizEygyq1vr-_VYkB16g2BOYolK-WMcop_1RxFVNcO_WHYAwRcFASvuF-AdXr5ViP1Ep
wvUn_L0enjdtjvlmlVIvqccAOpZGOht2U_luH-p9J1GWxmLv1AMj-wtGXhThNLFsBB-d7HU2v_D4RaQhOANjX-US-EvmBgaTwdJO29nxI5l3Fsi3II1pwXDkR8h_XRkMPSR8EDzZy3
f5OQc_pWAEjpetBxZCk0IOd15W25k8-Gr194sY_2oc_KLros1zL5u9gzFcZn6ZbDifsU9Yl0zAi1Zf1kqig0skxwLB3gNZ3TngWlpFlPYcYgNhTXC5FbA-xsgEBbaiRQLDYKi4Wg",
"expires_in":900,"refresh_expires_in":900,"refresh_token":"eyJhbGciOiJIUzI1NiIsInR5cCIgOiAiSldUIiwia2lkIiA6ICI1YzE0ZDE1MC00ZjA4LTRlMzctOWM
xZC03OTFmY2NjN2Y5ZWIifQ.eyJleHAiOjE2MjI5NTE0MzUsImlhdCI6MTYyMjk1MDUzNSwianRpIjoiYjY0ZmFhNzItZmViNS00NGY3LTlmZmItNGEyYzNlODU5YmYxIiwiaXNzIj
oiaHR0cDovL2xvY2FsaG9zdDo4ODg4L2F1dGgvcmVhbG1zL21hc3RlciIsImF1ZCI6Imh0dHA6Ly9sb2NhbGhvc3Q6ODg4OC9hdXRoL3JlYWxtcy9tYXN0ZXIiLCJzdWIiOiJkYTU0
ZTdkNy00MGE5LTQyNWEtODM4ZC1iM2U2NTU5ZjVjNzUiLCJ0eXAiOiJSZWZyZXNoIiwiYXpwIjoicGV0LWNsaW5pYyIsInNlc3Npb25fc3RhdGUiOiJhNzA4ZWU0OS04MDVlLTQ0Mj
ctODVhNS0xMmQzODY0MGNhZDYiLCJzY29wZSI6Im9wZW5pZCBlbWFpbCBwcm9maWxlIn0.3Ca51RpmFGPW8UPE7TJhOB213M2heH6Sk4V9edgi96o","token_type":"Bearer","
id_token":"eyJhbGciOiJSUzI1NiIsInR5cCIgOiAiSldUIiwia2lkIiA6ICIweThZU3VsUzYtZlYzWUNudFI2RjFuYUNVYVBUNkplT3k2amR5cUVXNnpnIn0.eyJleHAiOjE2MjI
5NTE0MzUsImlhdCI6MTYyMjk1MDUzNSwiYXV0aF90aW1lIjowLCJqdGkiOiI5NTI0YmZjYy04YTY3A3LTQ2ODgtYWU2Ni1jNjg2MzFlYTc4YzIiLCJpc3MiOiJodHRwOi8vbG9jYWxob
3N0Ojg4ODgvYXV0aC9yZWFsbXMvbWFzdGVyIiwiYXVkIjoicGV0LWNsaW5pYyIsInN1YiI6ImRhNTRlN1N2Q3LTQwYTktNDI1YS04MzhkLWIzZTY1NTlmN3M3NISsInR5cCI6IklEIiw
iYXpwIjoicGV0LWNsaW5pYyIsInNlc3Npb25fc3RhdGUiOiJhNzA4ZWU0OS04MDVlLTQ0MjctODVhNS0xMmQzODY0MGNhZDYiLCJhdF9oYXNoIjoiN3Z1RnJNT1dhSVR0bDRYVFJra
1RHQSIsImFjciI6IjEiLCJlbWFpbF92ZXJpZmllZCI6ZmFsc2UsInByZWZlcnJlZF91c2VybmFtZSI6ImFsaWNlIn0.Wg-UjATBMuRaoyIXaO65UTPD2BchvvuPa6PawOdbupKMrGt
Hh7a1-ah0WDnzIrvq2ZcQAdM9hLsh7O8kNLuyFHjoANXmto4UlmHxN64f_VRMj3yOrTpvqPMFjrQoWrJ4FojhCjJlgj4AnAyFHArEa1E0KmXXXRof2JCLKX0ZpKDDLVseXfKdy2Cev
J0jmFe_uLOJlZ74C8nAsYsP0OudrlQRrGe7MSM-XebRvyWpnSSMYnjopLTgv_0VZYWKVyCbSFe6R2R5EI7fW0JxD_ZF1Wd2y796SlYGxQIpjS37NfmlClFQCzf71buA_66klDxeXVr
Z2-pYrXiguG-vn2kprg","not-before-policy":0,"session_state":"a708ee49-805e-4427-85a5-12d38640cad6","scope":"openid email profile"}
zackdawood@Zacks-MacBook-Pro micronaut-petclinic %
```

Figure 11.18 – curl response JWT
```

10. Copy and format the JSON response using a JSON formatter. The `access_token` attribute value from the formatter response must be copied, and this value will be passed as a JWT, shown as follows:

Figure 11.19 – curl response JWT

11. Once you receive a response from `curl`, you can copy the `access_token` value. `access_token` can be passed to call the services in the request header, as shown in the following screenshot:

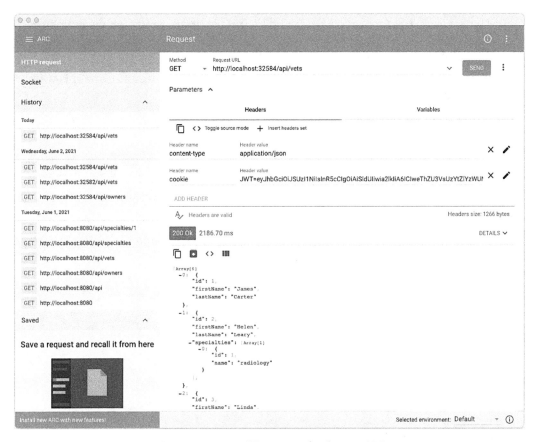

Figure 11.20 – REST response for the vets API

Now, we must test all the application URLs and their integration using the following links:

- **API gateway**: `http://localhost:32584/`.
- **Service discovery**: `http://localhost:8500/ui/dc1/services`. This will launch Consul as shown in the screenshot:

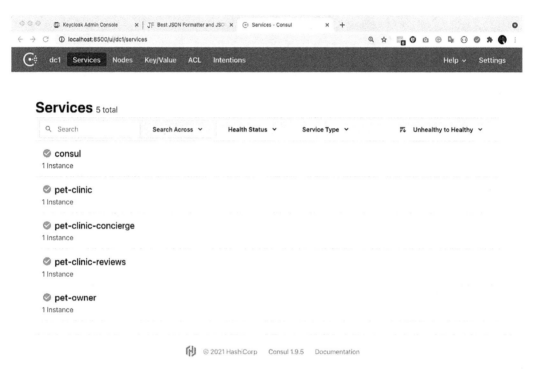

Figure 11.21 – REST response for the vets API

- **Distributed logging**: `http://localhost:5601/app/kibana`. This will launch the Kibana logging portal as shown in the following screenshot:

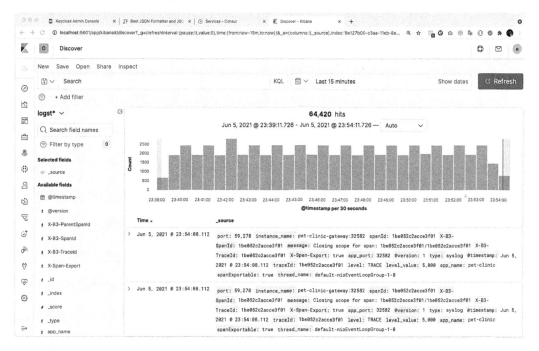

Figure 11.22 – Kibana logging portal

---

**Note**

The Kibana default username is `elastic` and password is `changeme`.

- **Distributed monitoring**: `http://localhost:3000/?orgId=1`. This will launch the Grafana monitoring tool. Detailed steps to configure the dashboard can be found in *Chapter 9, Distributed Logging, Tracing, and Monitoring*:

Figure 11.23 – Distributed monitoring using Grafana

---

**Note**

Prometheus, default username is `admin` and password is `pass`.

---

- **Distributed tracing**: `http://localhost:9411/zipkin/`. The following screenshot illustrates the Zipkin user interface for distributed tracing:

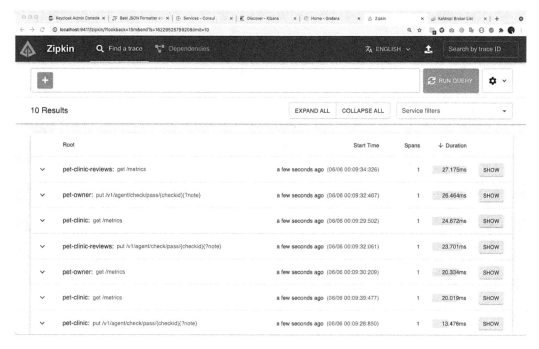

Figure 11.24 – Distributed tracing using Zipkin

- **IDP**: `http://localhost:8888/auth/` is the identity provider that will launch Keycloak which is the identity provider.

- **Kafka Kafdrop**: `http://localhost:9100/`. Here you can view the cluster with Kafdrop:

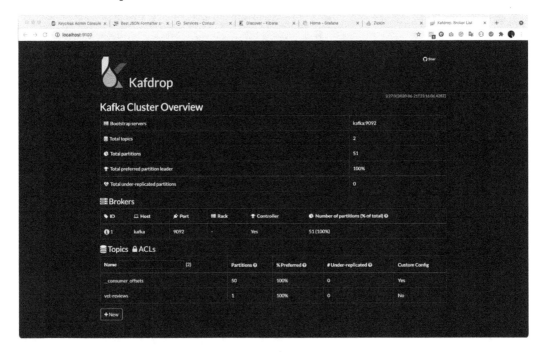

Figure 11.25 – Kafdrop Cluster view

Overall, we have tested all the features of the integration from the gateway to distributed monitoring, tracking, logging, and searching. With this, we have completed this chapter. In this section, we learned how to implement microservices in Micronaut while in production.

# Summary

In this chapter, we brought everything we'd learned about in the previous chapters together. Then, we dived into architecting enterprise-grade microservices. We also learned about scaling microservices, the different types of scaling, and their advantages.

This chapter enhanced your Micronaut microservices knowledge so that you can make production-ready applications. It has equipped you with all the necessary skills and expertise. We started the journey with the basics of microservices, and now we have the knowledge to create enterprise-grade microservices with Micronaut. We hope you enjoyed the journey of learning with us.

# Questions

1. Name a few factors you should consider when architecting microservices.
2. Name a few design patterns.
3. What factors should you consider during the design phase of microservices?
4. What factors should you consider during the development phase of microservices?
5. Name a few tools that are used in the deployment phase.
6. Name a few tools that are used in the manage and maintain phase.
7. What are the different types of scaling?
8. What type of scaling is used in a microservices architecture?

# Assessment

The answers for chapters 1 to 9 are mostly theoretical or practical. So it is enough to have read through them to answer the questions in those chapters.

## Chapter 10

1.  **Internet of Things** (**IoT**) is a network of devices or things. These things could be anything from a human wearing a health monitor, a pet wearing a geolocation sensor, a car with a tire pressure sensor, a television with voice/visual capabilities, or a smart speaker.

    We discussed this in the *Basics of IoT* section.

2.  Doorbells, locks, lightbulbs, speakers, televisions, healthcare systems, fitness systems, Google Home, Amazon Alexa, Apple Siri, Microsoft Cortana, and Samsung Bixby.

    We discussed this in the *Basics of IoT* section.

3.  Alexa skills are like apps, and you can enable or disable skills using the Alexa app for a specific device. Skills are voice-based Alexa capabilities.

    We discussed this in the *Basics of IoT* section.

4.  Intents capture events the end user wants to implement with their voice. An intent represents an action that is triggered by the user's spoken request. Intents in Alexa are specified in a JSON structure called intent schema. Built-in intents are **Cancel**, **Help**, **Stop**, **Navigate Home**, and **Fallback**. Some intents are basic, such as Help, and the skills should have a **Help** intent.

    We discussed this in the *Basics of intents* section.

5.  Java, Node.js, C#, or Go.

6.  We discussed this in the *Your first HelloWorld Alexa skill* section.

7. `LaunchRequestHandler`. `LaunchRequestHandler` is the first intent code that is triggered whenever a skill is launched.

   We discussed this in the *Your first HelloWorld Alexa skill* section.

8. `@Singleton` annotation. To change any Alexa Java handler to work with Micronaut, all we need to do is to add the `@Singleton` annotation, which is the `javax.inject.Singleton` annotation.

9. We discussed this in the *Integrating Micronaut with Alexa* section.

# Chapter 11

1. The following are a few factors to consider when architecting microservices are the following:

   - Does your team know about microservices?

   - Is your business mature enough to adopt microservices?

   - Do you have an Agile DevOps practice and infrastructure?

   - Do you have a scalable on-premise or cloud infrastructure?

   - Do you have support to use modern tools and technology?

   - Is your database ready to be decentralized?

   - Do you have support from all the stakeholders?

     We discussed this in the *Architecting enterprise microservices* section.

2. These are some design patterns:

   - Decomposition by business capability

   - Decomposition by domain/sub-domains

   - The API Gateway pattern

   - The chained microservices pattern

   - Database per service

   - The command query responsibility segregation pattern

   - The service discovery pattern

   - The circuit breaker pattern

   - The log aggregation pattern

We discussed this in the *Design* section.

3. The following are a few factors to consider during the design phase of microservices:

- Design patterns
- Security
- Authentication strategy
- Security rules
- Privilege-based access
- OAuth and SSL
- Privacy-specific standards such as HIPPA, GDPR, and PIPEDA
- Encryption at rest and encryption in transit
- A data retention strategy and log archival strategy
- Local country regulations
- Technology currency/versions

  We discussed this in the *Design* section.

4. The following are a few factors should you consider during the development phase of microservices:

- Always use the latest stable and supported version for development
- Leverage automation testing
- Test at various levels such as unit testing, service testing, and integration testing
- Use mocking and spying concepts during testing
- Emulate a real-world environment using containers or cloud infrastructure
- Synchronize development, QA, UAT, and PROD environments
- During development, think about backward compatibility for the microservice
- Have a separate database for each microservice to achieve the fullest potential

  We discussed this in the *Develop* section.

5. The following are a few tools that are used in the deployment phase:

- Containers
- Virtual machines

- The cloud
- Jib
- Jenkins
- DevOps Strategy and infrastructure to support Continuous Integration and Continuous Delivery

We discussed this in the *Deploy* section.

6. The following are a few tools that are used in the manage and maintain phase:

- Elasticsearch
- Logstash
- Kibana
- Prometheus
- Grafana
- Sonar

We discussed this in the *Manage and maintain* section.

7. The following are types of scaling:

- Vertical
- Horizontal
- Microservices horizontal

We discussed this in the *Scaling Micronaut* section.

8. The tool used is **Microservices Horizontal** (container/services-based scaling).

We discussed this in the *Scaling Micronaut* section.

Packt.com

Subscribe to our online digital library for full access to over 7,000 books and videos, as well as industry leading tools to help you plan your personal development and advance your career. For more information, please visit our website.

## Why subscribe?

- Spend less time learning and more time coding with practical eBooks and Videos from over 4,000 industry professionals

- Improve your learning with Skill Plans built especially for you

- Get a free eBook or video every month

- Fully searchable for easy access to vital information

- Copy and paste, print, and bookmark content

Did you know that Packt offers eBook versions of every book published, with PDF and ePub files available? You can upgrade to the eBook version at packt.com and as a print book customer, you are entitled to a discount on the eBook copy. Get in touch with us at customercare@packtpub.com for more details.

At www.packt.com, you can also read a collection of free technical articles, sign up for a range of free newsletters, and receive exclusive discounts and offers on Packt books and eBooks.

# Other Books You May Enjoy

If you enjoyed this book, you may be interested in these other books by Packt:

**Supercharge Your Applications with GraalVM**

A B Vijay Kumar

ISBN: 978-1-80056-490-9

- Gain a solid understanding of GraalVM and how it works under the hood
- Work with GraalVM's high performance optimizing compiler and see how it can be used in both JIT (just-in-time) and AOT (ahead-of-time) modes
- Get to grips with the various optimizations that GraalVM performs at runtime
- Use advanced tools to analyze and diagnose performance issues in the code
- Compile, embed, run, and interoperate between languages using Truffle on GraalVM
- Build optimum microservices using popular frameworks such as Micronaut and Quarkus to create cloud-native applications

**Microservices with Spring Boot and Spring Cloud – Second Edition**

Magnus Larsson

ISBN: 978-1-80107-297-7

- Build reactive microservices using Spring Boot

- Develop resilient and scalable microservices using Spring Cloud

- Use OAuth 2.1/OIDC and Spring Security to protect public APIs

- Implement Docker to bridge the gap between development, testing, and production

- Deploy and manage microservices with Kubernetes

- Apply Istio for improved security, observability, and traffic management

- Write and run manual and automated microservice tests with JUnit, testcontainers, Gradle, and bash

# Packt is searching for authors like you

If you're interested in becoming an author for Packt, please visit authors. packtpub.com and apply today. We have worked with thousands of developers and tech professionals, just like you, to help them share their insight with the global tech community. You can make a general application, apply for a specific hot topic that we are recruiting an author for, or submit your own idea.

# Share Your Thoughts

Now you've finished *Building Microservices with Micronaut*, we'd love to hear your thoughts! Scan the QR code below to go straight to the Amazon review page for this book and share your feedback or leave a review on the site that you purchased it from.

https://packt.link/r/1800564236

Your review is important to us and the tech community and will help us make sure we're delivering excellent quality content.

# Index

# V

# W